I0482661

# Disclaimer

The publisher of this book is by no way associated with the National Institute of Standards and Technology (NIST). The NIST did not publish this book. It was published by 50 page publications under the public domain license.

50 Page Publications.

**Book Title:** Benefits and Costs of Research: A Case Study of the NIST High Performance Concrete Program

**Book Author:** Jennifer F. Helgeson;

**Book Abstract:** This report provides an economic review of the National Institute of Standards and Technology High Performance Concrete (HYPERCON) Program from the period FY01 through FY09. The HYPERCON research program is designed to lower the cost of concrete performance prediction by developing and applying new measurement science including materials science understanding and performance prediction. The infrastructural nature of measurement science, its hard-to-visualize character, and the diffuse nature of its economic impacts makes them difficult to assess. The economic study of HYPERCON applies an innovative approach that uses surveys and case studies of the primary stakeholders within the industry to supplement more traditional quantitative success measures. Stakeholder survey responses are analyzed within the framework of grounded theory in order to build a cohesive understanding and assessment of the research program over time. The HYPERCON industry consortium known as the Virtual Cement and Concrete Testing Laboratory (VCCTL) is compared to three other major consortia in the cement/concrete industry to highlight its strengths and opportunities for improvement and identify consortia best practices. The economic study reports and summarizes its findings in order to help direct future HYPERCON Program investments in a way that meets the most important concrete performance prediction needs of U.S. industry in the most cost effective manner.

**Citation:** NIST TN - 1645

**Keywords:** Concrete performance prediction; economic impact assessment; grounded theory; Integrated Computational Materials Engineering; industry consortia; measurement science impacts; strategic basic research

NIST Technical Note 1645

U.S. Department of Commerce
National Institute of Standards and
Technology

Office of Applied Economics
Building and Fire Research Laboratory
Gaithersburg, Maryland 20899

# Benefits and Costs of Research: A Case Study of the NIST High Performance Concrete Program

Jennifer F. Helgeson

**U.S. Department of Commerce**
**National Institute of Standards and Technology**

Office of Applied Economics
Building and Fire Research Laboratory
Gaithersburg, Maryland 20899

# Benefits and Costs of Research: A Case Study of the NIST High Performance Concrete Program

Jennifer F. Helgeson

Sponsored by:

National Institute of Standards and Technology
Building and Fire Research Laboratory

**September 2009**

**U.S. DEPARTMENT OF COMMERCE**

Dr. Gary Locke, Secretary

**NATIONAL INSTITUTE OF STANDARDS AND TECHNOLOGY**

Dr. Patrick D. Gallagher, Deputy Director

Certain commercial entities, equipment, or materials may be identified in this document in order to describe an experimental procedure or concept adequately. Such identification is not intended to imply recommendation or endorsement by the National Institute of Standards and Technology, nor is it intended to imply that the entities, materials, or equipment are necessarily the best available for the purpose.

**National Institute of Standards and Technology Technical Note 1645**
**Natl. Inst. Stand. Technol. Tech. Note 1645, 105 pages (September 2009)**
**CODEN: NTNUE2**

# Abstract

This report provides an economic review of the National Institute of Standards and Technology High Performance Concrete (HYPERCON) Program from the period FY01 through FY09. The HYPERCON research program is designed to lower the cost of concrete performance prediction by developing and applying new measurement science including materials science understanding and performance prediction. The infrastructural nature of measurement science, its hard-to-visualize character, and the diffuse nature of its economic impacts makes them difficult to assess. The economic study of HYPERCON applies an innovative approach that uses surveys and case studies of the primary stakeholders within the industry to supplement more traditional quantitative success measures. Stakeholder survey responses are analyzed within the framework of "grounded theory" in order to build a cohesive understanding and assessment of the research program over time. The HYPERCON industry consortium known as the Virtual Cement and Concrete Testing Laboratory (VCCTL) is compared to three other major consortia in the cement/concrete industry to highlight its strengths and opportunities for improvement and identify consortia best practices. The economic study reports and summarizes its findings in order to help direct future HYPERCON Program investments in a way that meets the most important concrete performance prediction needs of U.S. industry in the most cost effective manner.

# Keywords

Concrete performance prediction; economic impact assessment; grounded theory; Integrated Computational Materials Engineering; industry consortia; measurement science impacts; strategic basic research

# Acknowledgements

The author wishes to thank all of those who contributed so many excellent suggestions for this report. Advice from members of the Office of Applied Economics (OAE) was integral to the success of the new assessment approach that this report took. Special thanks go to Dr. Robert E. Chapman and Ms. Barbara Lippiatt for their guidance throughout the development of the survey tools and their thorough review of this report. Thanks go to Dr. David Butry for his advice concerning reporting the findings. This study is the product of cooperation between the NIST Building and Fire Research Laboratory's (BFRL) OAE and its Inorganic Materials Group of the Materials and Construction Research Division. The author is grateful for the information that was provided by the Leader of the Inorganic Materials Group at the time of this study, Dr. Edward J. Garbozci, and other members of the Inorganic Materials Group, including: Dr. Dale E. Bentz, Dr. Jeffrey W. Bullard, Dr. Chiara Ferraris, Dr. Nicos Martys, Ms. Monyelle Mingo, Dr. Paul E. Stutzman, and Dr. Kenneth A. Snyder. Thanks go to Dr. Jonathan W. Martin, Chief of the Material and Construction Research Division, for his review of this report and the white paper leading to its development. Many thanks are extended to all of the cement/concrete industry stakeholder group members who took part in the various surveys. Thanks are also extended to current and past VCCTL members who were kind enough to take the time to complete surveys and interviews. The leadership of cement/concrete consortia made the case study comparison possible; thanks to Dr. Karen Scrivner of NanoCEM, Dr. Jason Weiss of ACBM, Dr. Surenda Shah of ACBM, and Dr. Jacques Marchand of SUMMA. Thanks are also due to Dr. Gregory C. Tassey of the NIST Program Office, Ms. Karen B. Perry of BFRL, and Dr. Allison L. Huang of the OAE.

# Contents

Abstract .................................................................................................................. iii

Acknowledgements ................................................................................................ v

List of Tables ........................................................................................................ ix

List of Figures ....................................................................................................... x

List of Acronyms .................................................................................................. xi

Executive Summary ............................................................................................ xiiii

1. Introduction .................................................................................................... 1

1.1 Cement/Concrete Industry Overview ............................................................. 1

1.2 Cement/Concrete Industry – Future Outlook .................................................. 3

1.3 The Strategic Basic Research Concept ............................................................ 4

1.4 Scope and Approach ....................................................................................... 6

2. BFRL HYPERCON Program ........................................................................ 7

2.1 HYPERCON Program Structure ..................................................................... 7

2.2 HYPERCON Strategic Basic Research: FY01 through FY09 .......................... 9

2.3 Virtual Concrete and Cement Testing Laboratory (VCCTL) ........................ 12

2.4 Supplemental HYPERCON Products ............................................................ 13

2.5  HYPERCON Stakeholders ........................................................................... 14

3. Economic Analysis Framework .................................................................. 19

3.1 Qualitative Approach: Overview .................................................................. 19

3.2 Survey Methods ........................................................................................... 20

3.3 Case Studies ................................................................................................ 21

3.4 Econometric Modeling: Qualitative Response Models .................................. 22

3.5 Grounded Theory ......................................................................................... 22

**4.    HYPERCON Economic Analysis: Approach** .................................................**25**

4.1 HYPERCON Quantitative Metrics ........................................................... 25

4.2 HYPERCON Qualitative Metrics ............................................................. 27
    4.2.1 HYPERCON Stakeholder Surveys .................................................... 27
    4.2.2 VCCTL Surveys and Consortia Comparisons ................................. 28

**5.    HYPERCON Economic Analysis: Data and Results** ................................**31**

5.1 HYPERCON Quantitative Metrics .......................................................... 31
    5.1.1 Standard Reference Material Usage .................................................. 31
    5.1.2 Guest Researchers / Academic Partnerships ..................................... 32
    5.1.3 Monograph Usage ............................................................................ 37
    5.1.4 ACBM/NIST Computer Modeling Workshop participation ............ 39
    5.1.5 SDO Membership ............................................................................. 42
    5.1.6 Other Agency and Scientific and Technical Research Services Funding ... 44
    5.1.7 VCCTL Consortium Membership .................................................... 45
    5.1.8 H-Factors .......................................................................................... 46

5.2 HYPERCON Qualitative Metrics ............................................................ 50
    5.2.1 HYPERCON Stakeholder Surveys: Overview ................................. 50
    5.2.2.Academia Survey Responses ............................................................ 50
    5.2.3 Aggregates Survey Responses .......................................................... 51
    5.2.4 Chemical Admixtures Survey Responses .......................................... 53
    5.2.5 State Departments of Transportation Survey Responses .................. 54
    5.2.6 Ready Mixed Survey Responses ....................................................... 56
    5.2.7 Internal Curing Research Program Responses .................................. 57
    5.2.8 Rheology Research Program Responses ............................................ 58
    5.2.9  X-ray Diffraction Research Program Responses ............................ 59
    5.2.10 Prescription to Performance Research Program Responses ............ 60
    5.2.11 VCCTL Research Program Responses ............................................ 62
    5.2.12 VCCTL Survey Responses ............................................................. 63
    5.2.13 Cement/Concrete Consortia Comparisons ..................................... 68

**6.    Summary** ..................................................................................................**77**

6.1 HYPERCON Retrospective Economic Impacts ...................................... 77

6.2 HYPERCON: Looking Forward .............................................................. 80

**References** .........................................................................................................**83**

# List of Tables

Table 1. Unit Sales of HYPERCON-Related SRM by SRM number: FY02-FY08 ..................... 33

Table 2. Unit Sales of HYPERCON-related SRMs: FY02-FY08 ................................................ 34

Table 3. Dollar Sales of HYPERCON-related SRMs: FY02-FY08 ............................................ 34

Table 4. HYPERCON Guest Researchers ................................................................................ 35

Table 5. HYPERCON Academic Collaborators ........................................................................ 35

Table 6. HYPERCON Academic Collaborations by Technical Focus Area ............................... 36

Table 7. ACBM/NIST Modeling Workshop Survey: Workshop Topics of Interest ................... 40

Table 8. SDO Committee Participation by HYPERCON Researchers ....................................... 42

Table 9. STRS and Leveraged R&D Funds per year ................................................................ 45

Table 10. VCCTL Consortium Membership as of February 2009 ............................................. 45

Table 11. H-Factors for HYPERCON and its Researchers ....................................................... 46

Table 12. Frequency of Citations to HYPERCON Research by Researcher: 2001-2008 ............ 50

Table 13. Past VCCTL Member Survey Response: Research Topics of Interest ....................... 66

# List of Figures

Figure 1. Flow of Information between HYPERCON Projects........................................................ 8

Figure 2. HYPERCON Component Projects and Outputs............................................................ 14

Figure 3. HYPERCON in the Grounded Theory Context ............................................................ 23

Figure 4. SDO and DOT Stakeholder Survey Responses: SRM Usage over Time...................... 34

Figure 5. Electronic Monograph Usage: FY02-FY08 ................................................................ 38

Figure 6. Electronic Monograph Survey Response: Significance of Impact................................ 39

Figure 7. HYPERCON Publications: 1970:2007........................................................................ 48

Figure 8. Citations to HYPERCON Publications: 1970-2009 .................................................... 49

Figure 9. HYPERCON Publications: 2000-2009 ...................................................................... 49

Figure 10. Citations to HYPERCON Publications: 2001-2009................................................... 49

Figure 11. Aggregates Survey Response: Applicability to Industry Stakeholders...................... 52

Figure 12. Aggregates Survey Response: Significance of HYPERCON Impact ......................... 53

Figure 13. State DoT Survey Response: Relevance of HYPERCON to DOTs............................ 55

Figure 14. State DoT Survey Response: Relevance of HYPERCON to Cement/Concrete Industry
.................................................................................................................................................... 55

Figure 15. Aggregates, State DoT, and Ready Mixed Survey Responses: Usage of HYPERCON
Internal Curing Research ........................................................................................................... 58

Figure 16. Aggregates, Admixtures, and SDO Survey Response: Usage of HYPERCON Applied
Rheology Research .................................................................................................................... 59

Figure 17. Aggregates and Admixtures Survey Response: Usage of HYPERCON Modeling
Rheology Research .................................................................................................................... 60

Figure 18. Admixture, Ready Mixed, and DoT Survey Responses: Usage of HYPERCON X-Ray
Diffraction Work........................................................................................................................ 61

Figure 19. Ready Mixed and DoT Survey Responses: Relevance of HYPERCON P2P Research
.................................................................................................................................................... 61

Figure 20. Aggregates, Admixtures, DoT, and Ready Mixed Survey Responses: Usage of
VCCTL Software........................................................................................................................ 62

Figure 21. VCCTL Members Survey Response: Significance of VCCTL Research Impact....... 64

# List of Acronyms

| | |
|---|---|
| ACBM | Center for Advanced Cement-Based Materials |
| ACI | American Concrete Institute |
| BFRL | Building and Fire Research Laboratory |
| DOE | U.S. Department of Energy |
| DOT | Department of Transportation |
| FHWA | Federal Highway Administration |
| GDP | Gross Domestic Product |
| HPC | High Performance Concrete |
| HYPERCON | High Performance Concrete |
| ICME | Integrated Computational Materials Engineering |
| IS | Impedance Spectroscopy |
| NIST | National Institute of Standards and Technology |
| NRC | National Research Council |
| NSB | National Science Board |
| NRMCA | National Ready-Mixed Association |
| OA | Other Agency |
| PCA | Portland Cement Association |
| QRM | Qualitative Response Model |
| REACT | Reduction of Early Age Cracking Today |
| SDO | Standard Development Organization |
| SRM | Standard Reference Material |
| STRS | Scientific and Technical Research Services |
| TIP | Technology Innovation Program |
| VCCTL | Virtual Cement and Concrete Testing Laboratory |
| XRD | X-ray Powder Diffraction |

## Executive Summary

This report provides an economic review of the National Institute of Standards and Technology High Performance Concrete (HYPERCON) Program from the period FY01 through FY09. The HYPERCON research program is designed to lower the cost of concrete performance prediction by developing and applying new measurement science including materials science understanding and performance prediction. The infrastructural nature of measurement science, its hard-to-visualize character, and the diffuse nature of its economic impacts makes them difficult to assess. The economic study of HYPERCON applies an innovative approach that uses surveys and case studies of the primary stakeholders within the industry to supplement more traditional quantitative success measures. Stakeholder survey responses are analyzed within a new conceptual framework in order to build a cohesive understanding and assessment of the research program over time. The HYPERCON industry consortium known as the Virtual Cement and Concrete Testing Laboratory (VCCTL) is compared to three other major consortia in the cement/concrete industry to highlight its strengths and opportunities for improvement and to identify consortia best practices. The report is intended to help direct future HYPERCON Program investments in a way that meets the most important concrete performance prediction needs of U.S. industry in the most cost effective manner.

The development of HYPERCON between FY01 and FY09 was tracked using the conceptual framework of *grounded theory,* a generalized way to view the evolution of a research program. The purpose of grounded theory is to develop theory about phenomena of interest, such as concrete performance. The theory (e.g. concrete performance models) needs to be grounded in observation. In a grounded theory approach to strategic basic research, the research begins with raising generative questions that guide the research process, but are intended to be neither static nor confining in nature. As the research team begins to gather data, *core theoretical concept(s)* are identified. Tentative linkages are developed between the theoretical core concepts and the data. These first steps can take years to complete. Subsequent research activities engage researchers in verification and summary; the effort tends to evolve towards one core concept that is central. Eventually, the research yields *conceptually dense theory* as new observation leads to refinement of tentative linkages and revisions in existing theory. This stage mirrors the concept of applied research to some extent; it is at this point that the core concept has been identified and fleshed out in detail and is disseminated to other industry players. The result of grounded theory is an extensively well-considered explanation for some phenomenon of interest, such as concrete performance.

Grounded theory provides a good context for the establishment of the qualitative effects of HYPERCON research. The strategic basic measurement science from HYPERCON and the program history of its VCCTL consortium are both at an early stage of development—identifying core theoretical concepts, gathering data, and identifying tentative linkages—precluding exclusive focus on quantitative indicators of economic impact. There is no viable quantitative metric that can alone track how the results of HYPERCON research that reaches industry, government, and academia is currently being used. Each individual stakeholder (e.g. company, individual) within a stakeholder category applies HYPERCON strategic basic research to their own applied research or production/use activities in a different manner. Thus, the

Economics of HYPERCON study interpreted emerging data on quantitative indicators through the lens of a qualitative assessment in the context of grounded theory.

Common themes emerged from this multi-pronged approach. As grounded theory suggests, HYPERCON technical focus areas have evolved over the period from FY01 through FY09, building upon research discoveries from period to period and adapting to changing industry needs. While fundamental measurement science issues remain in each of the five HYPERCON technical areas, HYPERCON's involvement in these areas since FY01 has generally evolved from providing the underlying measurement science toward addressing focused measurement issues enabling technical problem-solving and technology transfer. In FY09 HYPERCON was tied much more strongly and explicitly to national documents outlining the need for the kind of research it conducts. The research approach was directly identified as Integrated Computational Materials Engineering (ICME) for the first time. The Materials Characterization project was revamped to put the major emphasis on fly ash research. Modeling cement paste rheology with fly ash was assigned to the Rheology project. Additionally, long-range milestones were added to the Hydration Modeling project for fly ash and slag modeling.

Though they cannot be distilled into comparable (monetary) values, there are a number of quantitative metrics that were tracked as success indicators over the study period. The following table provides a summary of the major quantitative findings. Since FY09 data are preliminary estimates and quantitative data are largely unavailable for FY01, the table reports quantitative findings for the period FY02 through FY08.

| Quantitative Success Indicator | 2002 | 2008 | Total 2002 to 2008 | Average 2002 to 2008 | Percentage Growth 2002 to 2008 |
|---|---|---|---|---|---|
| Standard Reference Materials (SRMs) – Unit Sales | 959 | 1,096 | 6,764 | 966 | 14 % |
| SRMs – Dollar Sales | $ 118,318 | $ 177,460 | $ 977,532 | $ 139,647 | 50 % |
| Guest Researchers | 9 | 9 | 52 | 7 | 0 % |
| Electronic Monograph Use (number of different computers accessing) | 99,488 | 103,527 | 839,850 | 119,979 | 4 % |
| Other Agency (Leveraged) Funds | $ 101,500 | $ 457,600 | $ 2,186,300 | $ 312,329 | 350 % |
| VCCTL Consortium Fees (Leveraged Funds) | $ 147,400 | $ 368,100 | $ 2,871,600 | $ 410,229 | 150 % |
| Citations to HYPERCON Research (based on 2001-2008 publications) | 25 | 417 | 1,344 | 192 | 1,568 % |

Even in the face of economically trying times for the cement/concrete industry in the past few years, there is strong evidence of growth in most all quantitative indicators. Large fluctuations have taken place from year to year for some, as their totals and averages from FY02 through FY08 indicate. Interpretations of these results based on qualitative indicators, as determined through the survey process, follow.

Academic interest in HYPERCON appears to be quite strong and growing, with informal academic collaborations being an important venue for knowledge exchange. The ACBM/NIST Computer Modeling Workshop, which attracts a variety of stakeholders ranging from the academic to the industrial communities, has in recent years attracted significant interest among those specifically focused on computer modeling techniques, often times with no direct relationship to cement/concrete research. While this is a sign of HYPERCON's leading role in the ICME community at large, survey results indicate that the workshop is struggling to meet the growing, multi-faceted expectations of its participants. The Electronic Monograph, an inclusive record of HYPERCON research findings, is a relevant tool that serves many stakeholder groups' interests. The Monograph has proven to be one of the most effective (of many) channels for HYPERCON's impact within the academic realm. The HYPERCON Program's strong overall H-Factor from FY01 to FY09 points to the program's value and scholarly respect. Its publication base is large and growing steadily, with a clear upward trend in the number of attributed citations. One academic stakeholder expressed the sentiments of many when responding:

> *The NIST concrete group has put concrete at equal footing with other engineered materials, namely metals and ceramics. NIST thus largely contributes to reinventing concrete science and engineering as an academic discipline in the U.S. and worldwide.*

In the standards arena, there was an increasing trend in SRM sales from FY01 to FY09, particularly to foreign customers. HYPERCON's success in voluntary consensus standards development is more difficult to track, in part because of the very nature of standards development as a team exercise. Individuals and entities for the most part cannot claim "ownership" of a standard. While HYPERCON's clear contributions and direct impacts on standards from FY01 to FY09 are not evident, researchers' leadership positions on SDO committees are a positive indicator of future impact. Furthermore, SDO stakeholders generally support HYPERCON research and expect it will inform their future standards development needs.

As expected for a strategic basic research program, as the view shifts from academia to industry, there is a general trend for HYPERCON's retrospective impacts to become more diffuse. Stakeholder groups become more diverse in their levels of technical expertise and computer savvy and consequently, in how they view and plan to use HYPERCON research. Some of those interviewed are generally concerned with keeping up with the competition, while others are striving to position themselves as industry leaders. This diversity creates a challenge to HYPERCON to create a single mechanism for transfer of its research to industry and also affects the level of willingness for some companies to share their own strategic basic research activities with HYPERCON researchers. This is a common issue faced by any strategic basic research program and it is expected that as HYPERCON's grounded theory consolidates in the future and yields more tangible results, a more formalized solution to this problem may emerge.

The VCCTL is currently faced with an economy-driven decline in consortium membership, together with a consolidation towards exclusive membership among a lone stakeholder group (chemical admixtures). While consortium membership has fluctuated, past members generally found value in their participation. Past members support the idea of VCCTL research, but are finding it difficult to justify membership fees on a long-term basis. Among current and past consortium members and non-members alike, there appears to be interest in the VCCTL software, yet it does not align well with actual software use. Rather than a limitation in the VCCTL software itself (members report substantial improvement in the software interface), this likely highlights the naissance of the ICME approach, for which VCCTL is a leader. While many see a "significant potential application" for VCCTL in their future work, there is also a growing impatience for more tangible results.

The case studies of relevant cement/concrete consortia highlighted some areas in which VCCTL is a leader, as well as some comparative disadvantages in research structure. Particularly, the differences in consortia structure highlight the difference between strategic basic research and more applied research activities. The largest and most diverse cement/concrete consortium, NanoCEM, which is also the newest of the four reviewed, seems to benefit highly from a strict structure of research responsibilities and roles put in place by consortium management. Also, it is clear that the most widespread and quickest results occur when there is a wide pool of stakeholder groups represented by the consortium membership. In all consortia it is evident that members expect tangible results in return for financial contributions. This is much more difficult to achieve for a strategic basic research program, like VCCTL, than for consortia conducting applied research.

Across all stakeholders there is consistently a high respect for HYPERCON researchers and its program leader. Though, especially in industry, there is frustration over the speed at which findings are produced for HYPERCON research projects. This is a hallmark issue for strategic basic research programs. The usefulness and timing of their research findings are more difficult to anticipate, requiring flexibility in program planning.

HYPERCON is making progress in identifying tentative linkages among its core theoretical concepts. This is clear when questions about HYPERCON technical areas are asked to multiple stakeholder groups interested in a common area. Generally, it was found that while many of those surveyed are not aware of the details of HYPERCON research, those that were aware are confident that further HYPERCON research will be applicable to their work, particularly in those technical areas that are furthest along. ICME-related rheology is the exception, enjoying both good awareness and support among stakeholders. X-ray diffraction was found to be highly relevant to a range of stakeholders, from the beginning to the end of the cement/concrete industry supply chain. HYPERCON's P2P research—like P2P research in most areas—is struggling to define HYPERCON's role in fulfilling industry needs. Stakeholder groups were unanimous in looking forward to using the results of further HYPERCON research in their work. HYPERCON's progress in making linkages is also evident in its success at developing venues for bringing stakeholders together. It is important for stakeholders to have buy-in to HYPERCON's ultimate goal--in grounded theory's terms, conceptually dense theory for cement/concrete performance prediction. At this point, it is a matter of meeting HYPERCON expectations, defined internal to NIST and externally, and delivering the performance prediction models and tools its stakeholders clearly want.

While the nature of current HYPERCON research is still quite diffuse, there is great potential for significant future impact through multiple channels and serving multiple interests. There is no doubt that given the current economic climate there is a challenging road ahead for all cement/concrete strategic basic research. In the case of HYPERCON, success may depend on maintaining a delicate balance between being too ambitious and too weak in the promised outputs; it is key to avoid disenchanting those stakeholders that currently are actively engaged and highly supportive of HYPERCON's efforts. While stakeholder awareness of HYPERCON activities is not strong in some areas, this is not necessarily a place to focus on improvement in the immediate term. A more productive effort may be to develop and execute a vision that consolidates the overall program of research, so that when HYPERCON communicates with the currently uniformed, it can do so in terms of a compelling business case of mutual benefit to HYPERCON and U.S. cement/concrete industry.

Given the potential growth and importance of the industry in the coming years, NIST is poised to play an important role in cement and concrete research. The combination of sustained high asphalt prices, challenging U.S. economic conditions, and the need to improve and expand highway infrastructure creates favorable conditions for continued increases in concrete highway paving. Additionally, over the past decade advances have been made in the use of waste materials such as coal fly ash and blast furnace slag in cements, of crushed glass products in aggregates, and of recycled concrete. This is a trend that will continue and those surveyed expressed interest in understanding the chemistry and kinetics of concrete containing recyclables, especially through ICME interfaces. The combination of the current economic slowdown and environmental awareness contributes to a growing need to explore the value of waste stream materials to the concrete industry and construction industries as a whole. The use of such waste stream materials requires research, including measurement science and materials characterization, to be used effectively as a substitute for virgin materials in cement and concrete production.

There is a need for HYPERCON to consolidate its research projects into a program with a clear and succinct vision that is readily communicated. The grounded theory approach taken here shows that HYPERCON has proven its leadership in strategic basic research for cement/concrete. HYPERCON is now poised to capitalize on this leadership by communicating a compelling vision that could attract a wider user base to the VCCTL, both as potential members and collaborators. The VCCTL membership structure and governance processes may benefit from a review of the methods employed by other cement/concrete consortia.

HYPERCON research is applied throughout academia and industry in diverse and highly meaningful ways, which at this point can not be accurately tracked solely in monetary terms. This study has used grounded theory to assess HYPERCON performance from FY01 through FY09. Assessment of qualitative and quantitative indicators demonstrates that HYPERCON plays an important and potentially growing role in cement/concrete strategic basic research. Once the strategic basic research produced through HYPERCON permeates the industry and cost savings are apparent on an aggregate level, a review based on quantitative economic metrics should be possible. Then, the approach taken in this study could be enhanced with an econometric model based on two periods of data on HYPERCON quantitative and qualitative metrics to compare and use in the modeling.

# 1.    Introduction

Predicting concrete performance is a costly and difficult process. Concrete is a complex, multi-scale composite material, made using local materials.  These factors make concrete difficult to fully characterize, and performance is not easy to link to individual component materials or to combinations of component materials. The Building and Fire Research Laboratory (BFRL) at the National Institute of Standards and Technology (NIST) addresses these issues through the research efforts of the High Performance Concrete (HYPERCON) program. The HYPERCON research program was specifically designed to develop and apply new measurement science, including materials science understanding and performance prediction, to make concrete performance prediction possible and thus enable performance-based standards.

This study reports on HYPERCON's impact over the last eight years. It focuses primarily on impacts reported by key stakeholders within the cement/concrete industry.  The need for economic studies of NIST research programs is supported by the National Research Council's (NRC) biennial assessments of laboratory programs in the context of NIST's mission. One broad factor on which programs are assessed is the degree to which the Institute's measurement science and standards achieve their stated objectives and desired impact. In the analysis underlying this report, we take a similar approach by directly asking stakeholders about their satisfaction with key aspects of constituent HYPERCON projects. This retrospective study is meant to help direct future HYPERCON Program investments in a way that meets the most important concrete performance needs of U.S. industry in the most cost effective manner.

The Economics of HYPERCON study looks at the developments and changes to HYPERCON constituent projects and supplemental tools enabled by HYPERCON research from FY01 through FY09.  The economic assessment employs an approach using surveys, case studies, and a theoretical economic framework for a holistic assessment.  Qualitative and quantitative impact indicators based on application of this approach to HYPERCON are identified and evaluated.

In this section, the cement/concrete industry is described briefly and the scope and approach of this study is introduced.  The HYPERCON Program and associated projects are described in Section 2. Section 3 discusses the economic assessment framework and Section 4 its application to HYPERCON. Section 5 reports detailed data and results from applying the economic assessment framework to HYPERCON.  Section 6 concludes by summarizing the results and outlining the suggested "next steps" in the economic assessment of HYPERCON impacts.

1.1 Cement/Concrete Industry Overview

The cement[1] and concrete[2] industries represent primary inputs to the U.S. construction industry. Very little new construction can take place without the inputs of cement, and subsequently concrete.

---

[1] Cement is defined as a building material made by grinding calcined limestone and clay to a fine powder, which can be mixed with water and poured to set as a solid mass (cement paste) or used as an ingredient in making mortar or concrete.

The U.S. cement industry was valued with annual shipments of about $11.9 billion in 2007. Worldwide, the United States ranks third in cement production, following China and India. The United States has 116 cement plants operating in 38 states. The market share of companies in the cement sector is widely dispersed. The largest company produces 12.6 % of the cement industry total, and the top 5 producers account for 51.2 %. In the 1980s, foreign companies began to invest in the U.S. cement market. As of 2007, foreign companies, primarily European and Mexican, own 80.5 % of U.S. cement production capacity.

Cement consumption is seasonal; about two-thirds of U.S. cement consumption occurs from May to October.. There are major swings in the available inventory levels for cement and clinker[3] throughout the course of a year. Thus, cement producers tend to build up inventories during the winter months in anticipation of summer usage, which can be difficult to project in the current economic climate. Additionally, the cement industry is regional in nature, with 98 % of cement product shipped to customers by truck. Almost all of the cement produced is used to make concrete, worth at least $60 billion annually in the United States (USGS, 2007). About 1 % of the cement is consumed in the oil well drilling industry. Approximately 75 % to 80 % of all cement shipments are sent to ready-mix concrete operators in a given year. In 2007, plants shipped 13 % of manufactured cement to concrete product manufacturers (e.g. pre-cast concrete producers), 6 % directly to contractors (mostly road paving), and 3 % to building materials dealers. Section 2.5 provides further discussion of stakeholders in the cement and concrete industries.

The gaps in domestic production of cement and clinker are filled by foreign imports. The Portland Cement Association (PCA) estimates that U.S. cement plants had an average capacity utilization rate of 79 % in 2007. Even with a high plant operating capacity rate, 22.7 million metric tons of cement were imported to U.S. users. Eighty three percent of these imports are covered by five major countries: China, Canada, Columbia, Mexico, and the Republic of Korea. In 2007, the United States exported 1.6 million metric tons of cement to Canada; just exceeding 1 % of the total U.S. production.

In the past, reliance on foreign-produced cement has subjected U.S. users to volatile conditions related to the availability of foreign-produced cement as well as the dry bulk carriers needed for shipping. As a result, the cement industry has planned an aggressive capacity expansion program; by 2012, over 25 million metric tons of new capacity is scheduled to come online in the United States, an investment of more than $ 6 billion.

The performance of the U.S. concrete and cement industries is intimately connected to the fate of the U.S. construction industry through the mechanism of supply and demand. In 2007, the United States consumed 110.3 million metric tons of Portland cement; this is a 9.5 % decrease from 2006 consumption levels. The majority of this weakness can be attributed to reductions in

---

[2] Concrete is a hard, strong construction material consisting of sand, conglomerate gravel, pebbles, broken stone, or slag in a cement paste matrix.
[3] Cement clinkers are unfinished raw material (cinder lumps) formed by the heat processing of cement elements in a kiln.

residential construction. Residential construction spending was $829 billion in 2007, which is a 5.7 % decrease from 2006 levels. This consumption reduction is closely tied to the sub-prime issues in the residential sector, such as escalating mortgage rates on subprime loans, which are now spreading to the commercial sector, and will probably affect cement demand even more in the coming years. During 2007, spending on transportation infrastructure remained strong, funded in part by the SAFETEA-LU bill.[4]

1.2 Cement/Concrete Industry – Future Outlook

At the end of 2008, the Chief Economist at the PCA predicted: "the most pressure on the cement industry will materialize in 2009, when the trough level is reached, with an additional 5.5 % reduction in cement consumption."[5] The peak (2005)-to-trough (2009) decline is expected to reach 30 million metric tons, exceeding the rough recession years that the industry saw in 1974 and the early 1980s. In comparison to 2007 usage, U.S. national consumption of Portland cement was reduced by 15.3 % and masonry cement consumption went down by 28.8 % by 2009. The magnitude of the decline is exacerbated because at present the cement industry is engaged in the most intense expansion in the sector's history. Increases in production capacity are planned through 2012 and, though some of this increase may be ultimately reduced or unscheduled, PCA expects the cement market to return to peak levels (2005) by 2014.

The outlook for the cement and concrete industries in the immediate future is improved by proposed stimulus plan spending by the U.S. government. Sixty billion dollars of stimulus spending is planned for rebuilding U.S. infrastructure, of which $ 28 billion is intended for highways and bridges. The approximate use of Portland Cement in building highways and other public works is just over 35 % of total cement use (Perlman, 2009). The President and CEO of PCA sees this stimulus spending as "a once in a lifetime opportunity," but warns that there must be extreme care taken in choosing projects.[6] He warns very strongly to look for long-term quality fixes, rather than short-term quick fixes. Such an outlook comes at a period when there is a paradigm shift in the market; sharp increases in the price of asphalt and shortages in asphalt production products. Asphalt paving costs have increased 97 % during the past 5 years and more than 30 % during the past eighteen months, according to the Bureau of Labor Statistics. In terms of future costs, asphalt roads generally require maintenance every seven or eight years, whereas concrete roads last up to 30 years without serious repair (ibid.).

Thus, the combination of sustained high asphalt prices, challenging U.S. economic conditions, and the need to improve and expand highway infrastructure, creates favorable conditions for continued increases in concrete highway paving. More than 90 % of public sector concrete construction is spent at the state level. Thus, State Departments of Transportation (DOTs) will play a major role in determining the level of use of concrete in public works, and indirectly, the

---

[4] On August 10, 2005, President Bush signed into law the Safe, Accountable, Flexible, Efficient Transportation Equity Act: A Legacy for Users (SAFETEA-LU), which guarantees funding for highways, highway safety, and public transportation totaling $ 244.1 billion.
[5] Forecast by Edward Sullivan. Available: http://www.cement.org/newsroom/Spring08_Webcast.asp
[6] Forecast by Brian McCarthy. Available:
https://portlandcementevents.webex.com/mw0306l/mywebex/nbrDownload.do?siteurl=portlandcementevents

research applications in the cement/concrete field. Projected demographic changes will also direct the activities of DOTs. For example, PCA forecasts that by 2030, the U.S. will have 49 million more licensed drivers and total vehicle miles traveled is expected to increase by 49 %. In fact, the stimulus funds may only be the beginning of an expansion; according to the American Society of Civil Engineers, $ 1.6 trillion is needed during the next five years to repair and/or rebuild the existing highway infrastructure to acceptable conditions.

Efficiency gains in cement production over the past twenty years have been due primarily to automating production and significantly reducing small kiln production. In 2005, the cement production industry employed 16,877 workers, which is a 23 % reduction from 1985 levels. The average kiln produces about 74 % more cement than those used 20 years ago: 532,000 metric tons in 2006 versus 305,000 metric tons in 1986. Other efficiency increases can be attributed to phasing out energy-intensive wet kiln manufacturing[7] in favor of dry process cement manufacturing.[8]

Though cement production continues to gain efficiency, there is a great need for gains in efficiently testing cement to meet specifications of the users and to ensure the best possible service life of the product. HYPERCON's research activities for over a decade have addressed these issues, among others, and strived to achieve impact in an industry that has many players and competing interests.


1.3 The Strategic Basic Research Concept

HYPERCON's research objectives fall within a category which can be strictly classified as neither basic nor applied research. An accepted definition of basic research is: *research with an objective to gain fuller knowledge or understanding of the fundamental aspects of phenomena and of observable facts without specific applications toward processes or products in mind* (DOD, 2005). HYPERCON research activities are designed to enable applied research activities on the most pressing issues faced by concrete / cement producers and users. However, the complexity of concrete demands a fundamental approach (e.g. in Pasteur's quadrant).[9] Thus, the projects which constitute HYPERCON will be classified in this study as *Strategic basic research* activities. Strategic basic research is defined as: *fundamental research directed towards determining methods and knowledge relevant to a deep and significant issue within a certain field, which can then be refined through further applied research activities pertaining to more specific sub-problems.* Even though strategic basic research is defined by having a more specific end goal, its exploratory nature makes it a subcategory of basic research. The strategic basic research included under HYPERCON to date maps closely to generic technology for the

---

[7] Conventional wet process kilns are the oldest type of rotary kilns used to produce clinker. Conventional wet kiln technology has high heat consumption, requires larger kilns than the dry-kiln process, and produces large volumes of combustion gases and water vapor.

[8] Dry process kilns use dry raw materials. To improve the energy efficiency of the dry process, pre-heaters and pre-calciners have been introduced on newer kilns. No new wet kilns have been built in the U.S. since 1975.

[9] This reference is common with regards to the research of Louis Pasteur. Pasteur's basic research was motivated by practical objectives of improving industrial processes and public heath. It did lead directly to applications which saved the French silk and wine industries, improved the preservation of milk, and created effective vaccines. Thus, this is a prime example of basic research which contributes directly to applied research needs.

concrete and cement industry and infratechnologies produced. The following general discussion of basic research and the challenges involved with retrospective economic analyses thereof is relevant in this context.

Research and development activities in the United States represent a large enterprise, making up about 2.5 % of the nation's gross domestic product (NSB, 2008). The amount specifically devoted to basic knowledge-driven research is about 0.4 % of GDP. The impetus for continued federal support is summarized by the belief that "it is essential to recognize that technical advances depend on basic research in science…[basic science] is the wellspring of the technical innovations whose benefits are seen in economic growth…" (NRC, 2005). The inherent value of basic research to the economy is understood in general terms, but measurement of the direct impacts from basic research on technological innovation remains difficult to perform. This is attributed to the fact that there is a considerable lag in the transformation of basic research to applied research. Additionally, an applied research program is generally undertaken partially or wholly by private industry in order to transform basic research results into products. The combination of market risk assessments and estimates of technical risk complicates corporate R&D decisions and can increase time-lags from basic to applied research to products with direct economic impacts (Tassey, 1999).

A number of reports strive to link basic research value with quantifiable tangible outputs, but this rarely provides a holistic review of the research's true value. Pavitt (1996) demonstrates that linking basic research success with resulting patents is not an effective methodology. Patent records tend to cite previous patents in lieu of scientific journal articles; without thorough research it is impossible to determine whether the patent actually depended wholly on past basic research. Thus, patent counts and other *quantifiable* metrics may not wholly capture the significance of basic research. Martin and Tang (2006) cite scientific publications, citations, patents, licensing revenues, and spin-off companies as important, but imperfect/partial indicators of basic research success. Instead, they stress the importance of tracing the channels through which benefits of research flow into the economy, such as in the supply of skilled graduates and researchers, creations of new scientific instrumentation and methodologies, and enhancement of problem-solving capacity.

Evaluation of a strategic basic research program, especially in a public sector agency, is typically meant to oversee, improve, and make sense of the program. Ultimately it is the magnitude and structure of the program that serves to guide the analysts towards a certain style of analysis. When analyzing strategic basic research programs, which may not have had directly tractable economic impacts in the larger market, descriptive case studies, case studies, and sociometric and social network analysis is suggested (Ruegg and Feller, 2003).

The following section will present the scope, purpose, and methods of evaluation adopted in this retrospective study.

## 1.4 Scope and Approach

This study focuses on assessing outcomes and processes within the HYPERCON program during the period FY01 to FY09. The study was conducted during FY09, thus statistics comparable between years generally include data through FY08. Program evaluation looks at the impacts of a collection of projects over a fixed time period (ibid.). It is natural to compare outcomes in different periods in programs that involve multiple cycles in product and/or process innovation. HYPERCON, however, has been a single cycle program focused on "process innovation" – continual improvements in basic knowledge of concrete and cement. As such, direct comparison between periods of innovation is challenging. (Tassey, personal communication)

To date, one of three approaches has been utilized in isolation to determine economic benefits from publicly funded strategic research efforts: 1) surveys; 2) case studies; or 3) econometric models. Each approach has inherent biases and limitations. These approaches strongly consider qualitative as well as quantitative metrics. Opposed to case studies focused on economic estimation, the case studies we conduct are generally descriptive in nature. The use of descriptive case studies follows closely from suggestions for assessment of strategic basic research in the literature (Ruegg and Feller, 2003). Throughout this retrospective study a comparative method is employed over the qualitative indicators in the spirit of the "grounded theory" assessment framework. This methodology combines an analytical procedure of constant comparison, with an explicit coding procedure and a specific style of theory development (Glaser and Strauss, 1967). The Economics of HYPERCON project takes a holistic approach by considering all three of the aforementioned methods in tandem to avoid inherent biases of their use in isolation. This approach allows us to determine general HYPERCON trends throughout the study period.

Economic assessment of its programs helps BFRL managers meet the measurement science needs of the U.S. building and fire safety industries in a cost effective manner. Given the potential growth and size of the cement/concrete industry in the coming years, as discussed in Section 1.2, BFRL's concrete and cement research will become increasingly important. This study is intended to provide BFRL managers with a tool to review the activities of the HYPERCON program and associated projects and to develop a vision for future work to best meet the needs of the U.S. cement/concrete industry.

# 2. BFRL HYPERCON Program

BFRL's HYPERCON program, created in 1994 to develop measurement science enabling increased productivity in the U.S. cement and concrete industries, is the only federal research program focused on cement and concrete material science. The strategic basic research that the HYPERCON program undertook from FY01 to FY09 broadly seeks to address the lack of relevant performance-based standards, at least ones with strong connections to reality and sufficient predictive capability within the cement/concrete industry. Throughout this eight-year period there has been a well-documented need for such strategic basic research on concrete and cement (FHWA, 2006), which continues growing in the face of an estimated $1.6 trillion cost of revitalizing the concrete-dependent U.S. physical infrastructure (ASCE, 2005).

This section first discusses HYPERCON's program structure and its component parts. A description of the stakeholder groups associated with HYPERCON research follows, including a discussion of their research interests relative to HYPERCON research and products.

2.1 HYPERCON Program Structure

The HYPERCON Program conducts strategic basic research covering the following technical areas:

1. Materials characterization
2. Rheology/processing
3. Transport properties
4. Concrete technology
5. Computational materials science

Though specific project outputs,[10] goal outcomes,[11] and technical emphases have evolved over time, coverage of this spectrum of technical areas has remained fairly unchanged over the period FY01 to FY09. Information flow and exchange among these technical areas have also been maintained. As shown in Figure 1, there is both a top-down and bottom-up flow of information between technical areas. For example, the Modeling of Hydration project uses input data from the Characterization project. The Economics of HYPERCON project analyzes each technical area individually as well as the HYPERCON program as a whole.

---

[10] Program and project outputs are *major milestones* reflecting completion of specific and substantial activities.

[11] Program and project outcomes are *substantial, positive changes (i.e., deliverables)* directly enabled by, or due to, program/project outputs.

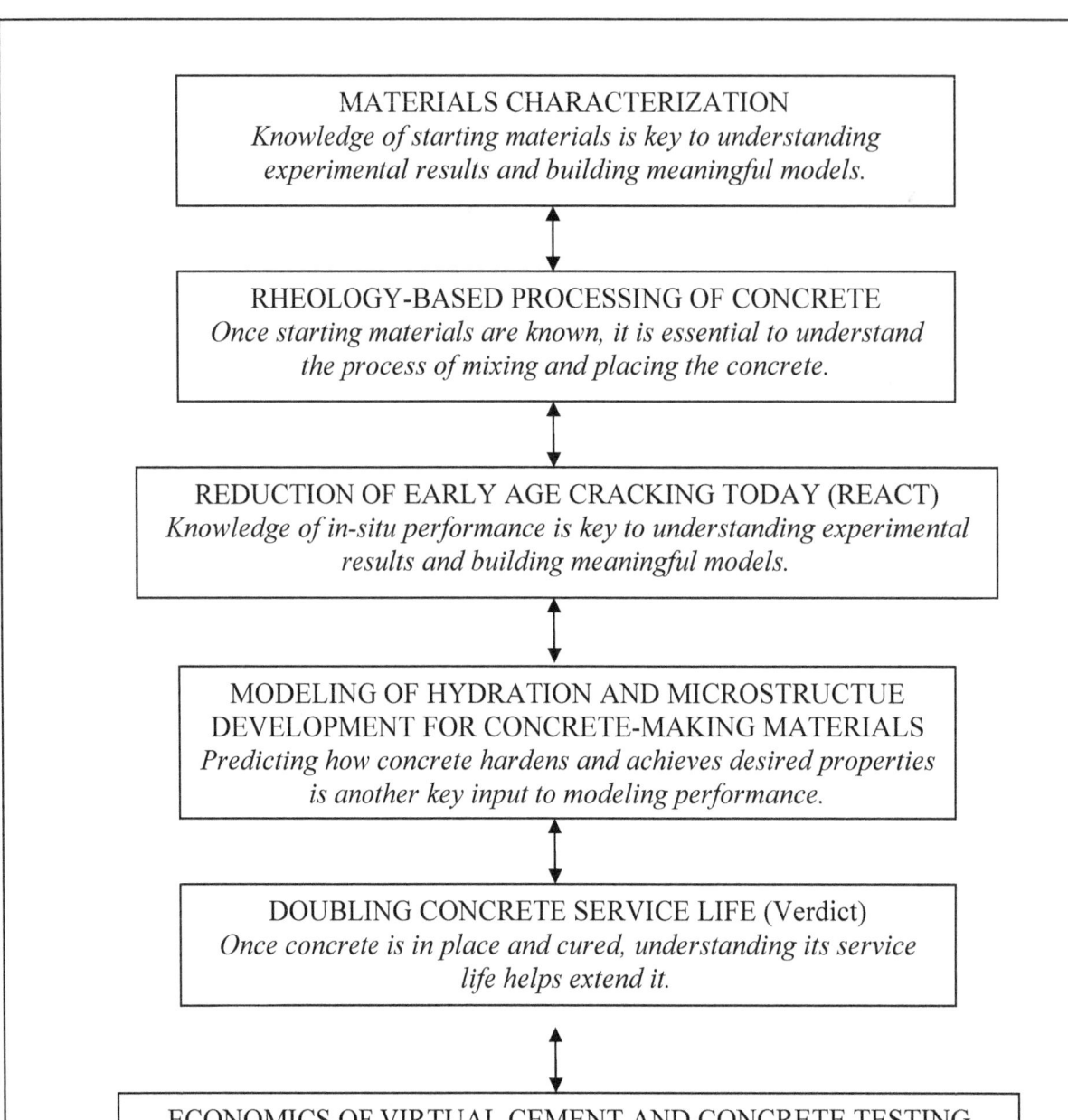

MATERIALS CHARACTERIZATION
*Knowledge of starting materials is key to understanding experimental results and building meaningful models.*

RHEOLOGY-BASED PROCESSING OF CONCRETE
*Once starting materials are known, it is essential to understand the process of mixing and placing the concrete.*

REDUCTION OF EARLY AGE CRACKING TODAY (REACT)
*Knowledge of in-situ performance is key to understanding experimental results and building meaningful models.*

MODELING OF HYDRATION AND MICROSTRUCTUE DEVELOPMENT FOR CONCRETE-MAKING MATERIALS
*Predicting how concrete hardens and achieves desired properties is another key input to modeling performance.*

DOUBLING CONCRETE SERVICE LIFE (Verdict)
*Once concrete is in place and cured, understanding its service life helps extend it.*

ECONOMICS OF VIRTUAL CEMENT AND CONCRETE TESTING
*Measure impact of HYPERCON basic research in performance prediction to help guide future research directions.*

*Figure 1. Flow of Information between HYPERCON Projects*

8

## 2.2 HYPERCON Strategic Basic Research: FY01 through FY09

The constituent projects in the HYPERCON research program have evolved over the period from FY01 through FY09, building upon research discoveries from period to period and adapting to industry needs.

### *Materials Characterization*

The Materials Characterization project seeks to develop quantitative materials characterization methods to accomplish more complete and accurate analyses of the phase composition and texture of cementitious materials. This will provide the basis for establishing new material-performance relationships for classes of cementitious materials that are produced today, which will become increasingly complex tomorrow. In 1996, an ASTM standard test method for point count analysis of Portland cement clinkers was developed. This method was used to create a set of Reference Material clinkers for phase analysis. At the same time, work was done on improving precision and accuracy in X-ray powder diffraction analysis of cements. The X-ray method was used to quantify phase proportions of the NIST Reference Clinkers and the data were merged with the microscopy data to generate certified values, upgrading the Reference Clinkers to Standard Reference Materials (SRMs). They continue to be in high demand with laboratories seeking to evaluate their instrumentation and analytical procedures for phase analysis.

Concurrent with this work, through participation in ASTM C 1.23 (Compositional Analysis), an X-ray powder diffraction (XRD) test method for Portland cement and clinker was developed. NIST project researchers co-authored the first standard test method and subsequently planned and coordinated an international inter-laboratory study to establish the precision and bias of the procedure. A major outcome of this study was a new ASTM XRD standard test method based using pattern analysis and crystal structure models, the first ASTM standard of this type.

This project is now using scanning electron microscopy to evaluate and quantify microstructural features of clinker, cement, and hardened Portland cement concretes. Project members are working to introduce this method into mainstream concrete petrography in ASTM C9 (Concrete) through demonstrations of evaluations of damaged concretes. Applying this method to tracking progressive degradation in sulfate-exposed concretes, a streamlined test procedure was developed that was submitted to ASTM and is currently under evaluation. This method was also applied to the latest SRM clinker (SRM 2686a) for the microscopy data set. Cements from the Cement and Concrete Reference Laboratory are now routinely imaged by this process to relate performance attributes to mineralogical and textural characteristics. In FY09, work on similar characterization of fly ash was begun.

### *Rheology/processing*

The establishment of this technical area of research predates the study start date of FY01. This research is focused around two "task areas." Task 1 involves developing a model for predicting rheological properties of high performance concrete (HPC) from the proposed mixture proportions and the flow properties of cement paste and mortar. The second task involves developing metrological methods for accurately measuring the rheology of cement paste, mortar, and concrete.

From FY02 onwards these two tasks were refined to allow a more in-depth understanding of rheology/processing of concrete. During the period FY02 to FY06, Task 1 involved validating the model using concrete, mortar, and cement paste rheological measurements in cooperation with University of Illinois and W.R. Grace researchers. The model was used to simulate flow in various applications, and a database linking coarse aggregate distribution with rheological properties was established.

Under Task 2, a number of alternative measurement techniques were explored and developed. Concrete rheometers were compared to initiate a potential standardization of the testing methodologies (completed FY04). This work was sponsored by ACI 236A and is reported in NISTIR 7154. Measurement methodologies were developed to address proper dosage and type of chemical admixtures for cement paste and mortar, as well as to aid in selection of supplementary cemenitious materials as a substitute for cement or sand.

In FY07 and FY08, the direction of the research under the two main task areas was further refined. Under Task 1 there was a strong refocus on developing a model to simulate the shear of concentrated suspensions and to validate it for concrete. Exploration of additional alternative measurement techniques for Task 2 were added in FY07. Additionally, in FY07, a NASA supercomputer was used to study the influence on concrete rheometry of particle shape, size, and distribution. In FY09, computer simulation will be used to determine the flow inside a rheometer and then to extract the viscosity of the material depending on the geometry. Also, the building of a unique instrumented pumping station will allow the measurements of critical parameters of the flow of grouts through pipes of different diameters and geometries.

### Transport properties

HYPERCON constituent projects have addressed the complex issues of fluid and ion transport in concrete from FY01 through FY09. In FY01, impedance spectroscopy (IS) was used to demonstrate that concrete conductivity can be measured accurately using the ubiquitous ASTM C 1202 apparatus. The demonstrated 5-minute conductivity test gives physical data more closely linked to the diffusive transport coefficient than the standard 6 hour test for total charge passed (ASTM C1202 Rapid Chloride Test). This work was published in the NIST Journal of Research. In FY01-FY02, there was a focus on the use of stable ceramic porous frits to demonstrate that the formation factor (ratio of the pore solution conductivity to the bulk sample conductivity) could be used as a meaningful transport parameter for diffusion in porous materials.

In FY03, a means of estimating the electrical conductivity of pore solution based on an arbitrary number of ionic species present was developed. This was a new development, as previous models only examined binary or ternary mixtures of species. Given that a speciation model could predict the pore solution composition, and the ASTM C 1202 test could quickly and accurately estimate bulk conductivity, the combination could serve as a means of estimating the formation factor. This work has led to development of an electro-diffusion equation that only requires the sample porosity and formation factor as transport parameters. "Negative Fickian diffusion" has been demonstrated in the laboratory and the behavior successfully predicted by the aforementioned equation. These advances were implemented in the project output, namely, the 4SIGHT computer model for performance assessment. The technical advances were also incorporated into the commercial STADIUM computer code (SIMCO Technologies, Inc.), which is being further developed for the Department of Energy (DOE)-NRC Cement Barriers Partnership, a project funded by the US Department of Energy of which HYPERCON is a partner.

### *Concrete Technology (Curing)*

The Reducing Early-Age Cracking Today (REACT) project was initiated in FY05 and has undergone two name changes, of which REACT is the current title. In February 2005, the research resulted in publication of a methodology for mixture proportioning for internal curing in *Concrete International*, and the significance of this contribution has been recognized through the receipt of the ACI 2007 Wason Medal for Materials Research. In June 2005, ASTM approved a NIST-drafted standard test method based on this work, now known as the ASTM C1608 "Test Method for Chemical Shrinkage of Hydraulic Cement Paste." In FY06, in collaboration with Northeast Solite and the Pennsylvania State University, REACT researchers conducted industry-funded experiments that utilize X-ray microtomography to directly observe water movement from pre-wetted lightweight aggregates to the surrounding cement paste during the internal curing of a mortar. In FY07, the principal investigator, Dale Bentz, presented an invited keynote lecture on "Early-Age Properties" at the 12th International Congress on the Chemistry of Cement in Montreal, Canada. In FY08, the resulting paper was published in the February 2008 issue of *Cement and Concrete Research* and led to an invited paper on "Reducing Early-Age Cracking Today (REACT) that was published in the June 2008 issue of *Concrete Plant International*. In FY09, in collaboration with Purdue University researchers, the team is slated to initiate a REACT consortium effort that will address early-age cracking issues, as identified and funded by the U.S. concrete industry.

### *Computational materials science*

The first project, Simulation of Concrete Performance, has been part of HYPERCON for 10 years or more and was completed in FY08. Its purpose was to develop new algorithms that accurately simulate concrete performance aspects and to incorporate these algorithms into publicly available software or into computational materials science code. Both methods of distribution have been used previously for conductivity/diffusivity and mechanical property algorithms. In the last few years, the objectives of this project have turned towards measuring and analyzing particle shape (e.g. cement, sand, and gravel). This process has allowed realistic particle shapes to be incorporated into all models that are used in HYPERCON computational materials engineering software.

The second project under this theme is the Cement Hydration project. From FY02 to FY04, research in this project was focused on generalizing the existing hydration model, CEMHYD3D, and upgrading the software user interface. Beginning in FY04, development of a new model of cement paste hydration, HydratiCA, was initiated and that work has continued to the present. In FY06-FY07, the HydratiCA model was extended to provide insight into the mechanisms of hydration of tricalcium silicate, the most important component of cement. Since FY07, the cure chemistry of cement has been more realistically modeled. In FY09, work will begin on incorporating pozzolans, such as fly ash and slag, into the model and on developing new models for predicting later-age hydration properties of cement and concrete.

While fundamental measurement science issues remain in each of the five HYPERCON technical areas, HYPERCON's contributions in these areas since FY01 has generally evolved from providing the underlying measurement science toward addressing focused measurement issues enabling technical problem-solving and technology transfer. To this point, the objective of the HYPERCON program though 2013 is: "to develop and implement the measurement science foundation that will give the concrete industry and state and federal government agencies the predictive capacity upon which they can base the use of performance-based standards and specifications in the key areas of diffusion-based life prediction, curing of mixed cemenitious systems, avoidance of early-age cracking, concrete placement via pumping, and greater use of fly ash." This evolution in HYPERCON's focus is demonstrated through the research activities of the computational materials engineering consortium, VCCTL, and through supplemental HYPERCON products, discussed below.

2.3 Virtual Concrete and Cement Testing Laboratory (VCCTL)

In FY01, HYPERCON began the Virtual Cement and Concrete Testing Laboratory (VCCTL) component project, which has grown into a government-industry consortium of concrete and cement industry stakeholders. VCCTL draws on certain HYPERCON strategic basic research project outcomes with the objective of providing industry with a virtual testing tool for reducing the number of physical concrete tests needed. Thus, VCCTL has served as a bridging mechanism from HYPERCON strategic basic research towards applications in industry. VCCTL has found use in the cement and concrete industries, with applications ranging from research and development, to mixture proportioning and troubleshooting.

The VCCTL Consortium aims to enable performance-based standards for cement and cemenitious products by overcoming technical barriers to performance prediction. Since consortia members usually must undertake additional, in-house R&D to fully benefit from the outcomes of R&D alliances such as VCCTL, its economic impact is difficult to evaluate based on tangible results alone. Yet there is a documented industry need for such consortia. The materials industry requires prediction capabilities for complex modern materials, among which concrete is arguably one of the most complex. Recently, the National Research Council has called for Integrated Computational Materials Engineering (ICME) models for all classes and applications of materials." (NRC, 2008 ) Specifically, the NRC has recommended that "U.S. industry identify high-priority foundational engineering problems that could be addressed by ICME, establish consortia, and secure resources for implementation of ICME into the integrated product development process." In the future, large scale advances enabled by strong ICME programs are expected to offer stimulation of economic development within the United States. The VCCTL government-industry consortium is a trail blazer with regards to ICME technology through its work on the VCCTL computational materials engineering software, rheology modeling, and hydration modeling since 2000. Members of 12 leading companies and organizations in the concrete industry have been paying members of the VCCTL consortium at one time or another since 2001, actively collaborating with and supporting HYPERCON research activities.

The NRC recognizes the technical challenges associated with adoption of ICME approaches. The fundamental technical difficulty highlighted is that the materials properties that are essential for design and manufacturing involve a multitude of physical phenomena, and accurately capturing their representation in models requires spanning many orders of magnitude in length scale and time. The NRC concludes that experimental efforts to calibrate both empirical and theoretical models and validate the ICME capability are paramount challenges to the general approach. Addressing this important concern for the cement and concrete community is precisely the challenge upon which the VCCTL was initially formulated and has sought to address in the last eight years.

A discussion of VCCTL in the context of other concrete and cement research consortia is provided in Section 5.2.3. Additionally, analysis of VCCTL research activities by past and present VCCTL consortium members was a main focus of survey activities, the findings of which are presented in Section 5.2.2.

## 2.4 Supplemental HYPERCON Products

HYPERCON project outcomes support supplemental products, which extend the HYPERCON knowledge base to industry and academics involved with cement and concrete production and use. The two most notable such projects, the Center for Advanced Cement-Based Materials (ACBM)/NIST Modeling Workshop and the Electronic Monograph: Measuring and Modeling the Structure and Properties of Cement-based Materials, are discussed below.

### *Electronic Monograph*
The Electronic Monograph[12] began in 1997 and records the work that has been developed thus far in the computational material science of concrete. The major text source is NIST published papers regarding concrete research activities. Because HYPERCON has served as a locus for computer modeling of the microstructure and properties of concrete since 1994, the majority of information contained in Monograph focuses on HYPERCON research output and research done in collaboration between HYPERCON and outside researchers and academics.

### *ACBM/NIST Computer Modeling Workshop*
The ACBM/NIST Computer Modeling Workshop, which is co-sponsored by ASTM committees C01 (cement) and C09 (concrete), was established in 1990. The workshop lectures cover topics in computational and experimental materials science of concrete, including simulation of microstructural development and prediction of physical properties. The workshop format allows for a mixture of longer tutorial sessions and short technical talks by the participants. For the 20 workshops to date, the total attendance was 567 individuals (2/3 from U.S. institutions and 1/3 from foreign institutions) representing 23 countries.[13] These attendees are distributed among academia, industry, and government as follows:

---

[12] The electronic Monograph can be accessed at: http://ciks.cbt.nist.gov/monograph/
[13] Australia, Belgium, Brazil, Canada, Chile, Czech Republic, Denmark, France, Germany, Italy, Japan, Korea, Mexico, Netherlands, Norway, Poland, Portugal, Singapore, South Africa, Spain, Sweden, Switzerland, United Kingdom

- Doctoral and Post-doctoral students: 277
- University Faculty: 102
- Industrial researchers: 116
- U.S. Government researchers: 72

In the survey portion of the Economics of HYPERCON study, the distribution of respondents was roughly proportional to this reported distribution.

HYPERCON is a diverse and extensive program made up of many constituent projects and products, as described above. Figure 2 provides a simplified schematic of these HYPERCON component projects and products. The items designates with a "Q" in their title indicate HYPERCON products. Items designated with a "P" in their title designate some of the HYPERCON related project outputs.

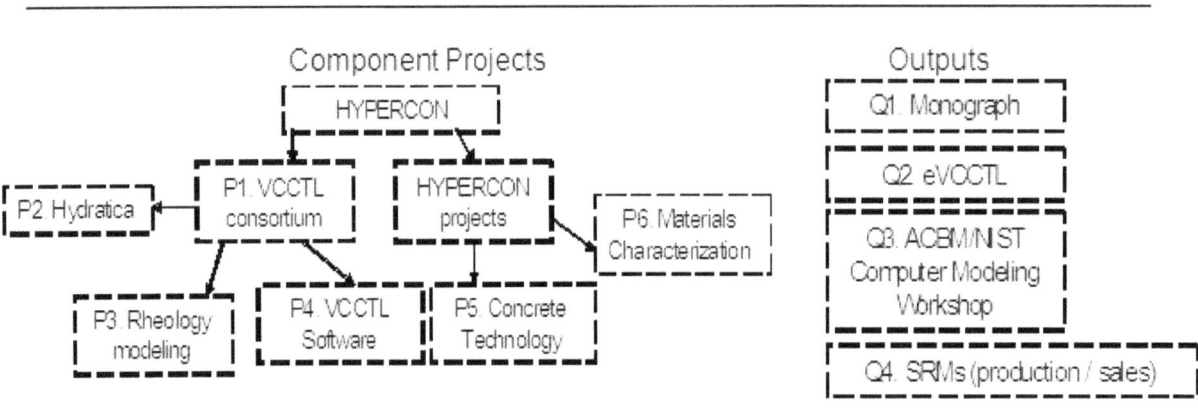

*Figure 2. HYPERCON Component Projects and Outputs*

To provide context for an economic assessment of the HYPERCON Program, its cement and concrete industry stakeholders are described in the next subsection.

## 2.5 HYPERCON Stakeholders

The cement/concrete industry is widespread and interests among stakeholders vary greatly, as does market share. These factors pose challenges to HYPERCON, which is specifically tasked with developing a strategic basic research agenda that "meets the most important concrete performance predictions needs of U.S. industry in the most cost-effective manner." The role of each stakeholder group in the cement and concrete industries is described below. A brief discussion and schematic representation of the interaction between the stakeholder groups and their use of HYPERCON outputs follows.

### *Cement*
Refer to Section 1.1 for a detailed discussion of the U.S. cement market and its economic interactions within the greater context of the U.S. construction market and concrete usage.

*Aggregates*
Aggregates are granular materials (e.g. sand, gravel) which in combination with water and cement, produce concrete. Aggregates account for 60 to 75 percent of the total volume of concrete.

Excavation of aggregates is a capital-intensive operation with large earth-moving equipment, belt conveyors, and crushing and separating machines being needed. According to the USGS Mineral Industry Surveys (June 2008), an estimated 414 million metric tons of aggregates was produced and shipped for consumption in the U.S. during the first quarter of 2008. This is a decrease of 16 % compared with that of the same period in 2007. Production-for-consumption of aggregates decreased in 30 of the 47 states covered by the survey. The same pressures facing the U.S. construction industry at present, discussed in Section 1.1, spill over into aggregate production. The market trend for aggregate companies at present is the purchase and consolidation of small companies by global corporations.

There has been advances in the use of recycled aggregates in concrete production over the past decade or so. The combination of an economic slowdown and environmental awareness contributes to a growing need to explore their value to the concrete and construction industries as a whole. Some notable examples are the increased use of recycled concrete and the use of crushed glass products as concrete aggregates. The use of recycled aggregates requires research and effective measurement science characterization to be used effectively as a substitute for natural aggregates in concrete production.

*Admixtures*
Admixtures are ingredients other than water, aggregates, and cement that are added to the concrete batch before mixing. This is a large market due to the performance enhancement that chemical admixtures lend to concrete. Market share of this industry sector is large per company; there is a handful of major producers due to the level of research and high capital costs involved. Many of the leading companies offer admixtures for concrete but also other concrete enhancement products, such as corrosion inhibition systems.

*Ready-Mixed Concrete*
Ready-mixed concrete refers to concrete that is batched for delivery from a central plant rather than prepared and mixed on-site. The Business Administration Committee of the National Ready-Mixed Concrete Association (NRMCA) administers the annual Industry Data Survey annually. to provide a benchmarking tool for companies in the industry. There were 172 respondents to the 2007 survey, which found average sales for them to be $ 72,742,605, or 866,414 cubic yards (662,421 cubic meters). This is a reduction of 1.7 % and 11 %, respectively, from 2006 levels.

*Pre-cast Concrete*
Precast concrete is concrete cast in a reusable mold that is cured in a controlled environment, then transported to the construction site for installation. Pre-cast concrete producers account for about 25 % of the total market of concrete produced.

### State Departments of Transportation

The transportation infrastructure in the United States is represented by the 50 state departments of transportation (DOTs). One of the main applications of concrete is in road construction among airport, railway, and bridge construction. State DOTs plan, authorize, and pay for transportation infrastructure within their own state boundaries. DOT controls this infrastructure on Federal lands, but everywhere else, it is owned by states and counties. They fund and do research, but this research tends to be short-term and empirical, solving immediate problems. They do fund the National Cooperative Highway Research Program, which has some projects that tend to be more long-range and basic in nature.

### Designers / Contractors

Designers specify concrete properties in their designs so as to have a workable structure. They are mainly interested in concrete performance, and not as much in the details of concrete mixtures. The designer/contractor stakeholder group has less incentive to be concerned with long-term concrete performance than other stakeholder groups, e.g. state DOTs, which maintain ownership of the built structure over its lifetime.

### Testing Laboratories

The main purpose of testing laboratories, in the context of concrete materials, is to supply data on how a contractor's materials score on various standard tests, whether ASTM, International, the American Association of State and Highway Transportation Officials (AASHTO), or some state or locality-specific test. A typical construction project, especially those done by state DOTs, includes testing requirements among the specifications for contractors to meet.

### Standard Developing Organizations (SDOs)

An SDO is defined as any entity whose primary activities are developing and maintaining standards that address the interests of a significant number of users outside of the SDO itself. HYPERCON activities are relevant to sub-committees of three major SDOs in which concrete industry stakeholders are active: ASTM International, the American Concrete Institute (ACI), and AASHTO.

These SDOs are interested in the work that HYPERCON does in bridging the gap between basic material science data,which can be used in models for performance prediction, and practical industry guidance.

### Academia

Almost all the academic work in concrete materials lies in civil engineering departments, where a few faculty out of many specialize in civil engineering materials. There is some work scattered across the United States in chemical engineering and materials science departments. ACBM, in its early years, was successful in getting faculty from materials science and physics departments interested in cement-based materials, but that interest has diminished over the last 20 years. Most of the research funding coming to civil engineering materials faculty comes from state DOTs or the National Cooperative Highway Research Program, or from industry.

***Nuclear Facilities***

Nuclear facilities have a stake in the cement and concrete industries given the surge of interest in nuclear power in the United States (Holton, 2005). The role of concrete for nuclear structures (e.g., power plants, spent fuel pools, barriers for waste containment) is the same as in other buildings and facilities, but there is an enhanced emphasis on quality and durability. New nuclear power plants will be designed with a 120-year service life, and concrete barriers for nuclear waste must be designed to last hundreds of years. HYPERCON may play a significant role in updating the concrete standards needed for new generation nuclear reactors. There is also a role for cement in nuclear applications apart from buildings. Grouts (mortars) and cements can be used to fill emptied liquid waste tanks or used to build waste forms that seal radioactive waste into containers of various kinds.

The specific needs of each stakeholder group within the cement/concrete industry are addressed by a selection of HYPERCON constituent projects and activities.

The surveys developed and discussed throughout this report are highly focused around the needs of identified stakeholders. Due to the structure of research undertaken by various actors in the cement/concrete industry, some stakeholders have a more active role than others in the research topics undertaken by HYPERCON over time. In this report, pre-cast concrete producers, nuclear stakeholders, and designer/contractors have not been included in the individual stakeholder surveys.

# 3. Economic Analysis Framework

This section provides a general introduction to the methods employed in the economic analysis framework for assessing the impacts of HYPERCON from FY01 through FY09. The fact that HYPERCON has historically taken on strategic basic research, paired with the fact that cement/concrete industry stakeholders vary so greatly in their applied research needs and actual ability to conduct applied research, makes it unrealistic to use a traditional benefit-cost framework for economic analysis. Namely, there is no viable quantitative metric that alone can track how the results of HYPERCON research impact industry, government, and academia. Each individual stakeholder (e.g. company, individual) within a stakeholder category applies HYPERCON strategic basic research to its own applied research or production/use activities in a different manner.

Counterfactual analyses, which seek understanding of a phenomenon through analyzing the situation under circumstances which run contrary to reality, are becoming a common method to assess impacts of a change in the structure of and interactions within an industry. Counterfactual studies hinge on the specification of synthetic scenarios, such as assuming how the entire concrete industry would now be different in the absence of HYPERCON research activities. Due to the differentiation and scope of the concrete industry, and the often isolated pockets of stakeholders HYPERCON serves, it is most effective to focus on impacts on a single actor basis (e.g. surveys). Use of a counterfactual approach will likely be relevant to future HYPERCON impact; however, once planned projects focused clearly on applied research are completed.

After a review of these and other modeling issues, a practical approach was developed that employs a combination of quantitative and qualitative methods. These methods will be applied in the context of *Grounded Theory* to appropriately accommodate the HYPERCON program and the unique structure of the concrete industry it serves. Because they are less well understood than quantitative measures, the following subsections provide background on qualitative measures and Grounded Theory.


3.1 Qualitative Approach: Overview

Qualitative analysis has been well established through sociological studies that strive to determine motivation and understanding underlying response patterns. Before undertaking this analysis of HYPERCON, the application of qualitative metrics in retrospective and prospective studies conducted by the NIST Technology Innovation Program (TIP, formerly the Advanced Technology Program), were considered. In its "Toolkit for Evaluating Public R&D Investment," there is a thorough review of a range of qualitative evaluation methods.

Stevens (1946) introduced a widely accepted definition of measurement: "the assignment of numbers to objects or events according to a rule." He proposed four "levels" of measurement: 1) nominal (categorical, discrete); 2) ordinal; 3) interval; and 4) ratio. There continues to be debate about the merits of classifications, particularly in the cases of nominal and ordinal classifications (Michell, 1986). Qualitative data is data that give non-numerical information; ordinal data is data about order or rank on a scale; and metric data is obtained from direct measurement of quantities.

In a statistics sense, quantitative variables take on numerical values and allow traceable, *objective* measurements. Yet, attribute coding schemes also allow "quantification" of qualitative data (Bateman, 2006). The kinds of descriptive statistics and significance tests applicable to such coding schemes depend on the level of *measurement* of the variables concerned. The manner by which the qualitative data was first obtained also plays a pivotal role in how it is most effectively analyzed.

The next three subsections describe modes of qualitative data collection considered in this study and their relevance to the analysis of HYPERCON.

3.2 Survey Methods

A well designed survey instrument is required to gather meaningful qualitative data. Such a survey instrument should clearly communicate relevant information and present questions in an unambiguous manner. Other qualitative research methods, such as focus groups, can facilitate decisions about what information to include in the survey. Bateman (2006) recognizes five main steps in development of the survey tool: 1) identifying desired measures; 2) writing survey questions; 3) balancing open- and close-ended questions; 4) combining with existing data-sets; and 5) question order and formatting. Surveys tend to use stated preference methods to obtain the desired measures. Stated preference relies on data from carefully worded survey questions asking individuals or enterprises what choice they would make for alternative levels of an amenity. The survey tools developed in this research used the stated preference method.

When "quantification" of qualitative measures is a priority, it is important to base survey responses on a well-explained, consistent scale. A Likert scale is a psychometric scale commonly used in questionnaires, and it was used in this study. This is a bipolar scaling method, which measures the extent of positive or negative response to a statement. A Likert item is a statement to which the respondent is asked to evaluate his or her level of agreement or disagreement according to a surveyor-defined subjective or objective criterion. Often five ordered response levels are used ("strongly disagree," "disagree," "neither agree nor disagree," "agree," "strongly agree"). This scaling is made more specific when the surveyor provides detailed descriptions of the response choices, such as "what it means" to "disagree." These responses can then be attribute coded (e.g. "strongly agree"=5, "disagree"=2). Ultimately, Likert scales are desirable because each response item may be analyzed separately, while mathematical summation of response categories can also be used to create a score for a group of items.

There is considerable literature concerning the merits of various modes of survey transmission, especially with regards to response quality. The low-cost and time efficiency of on-line survey delivery allows a greater pool of potential respondents to be initially contacted, thus increasing sample heterogeneity. Dohman (2005) determined that interviews conducted in-person differed negligibly from posted-response surveys. There are advantages to follow-up in person or telephone interviews, namely that the interviewer can record data on the impetus behind individual or enterprise survey responses.

All surveys in this study were scripted in and disseminated through the online application, Survey Monkey. In some cases follow-up phone and email contact was established. To ensure anonymity, responses were not traceable back to their source. Thus, some semi-structured phone and in-person interviews were conducted, but not specifically as follow-up to responses to the online surveys.

The stakeholder groups within the cement/concrete industry with which NIST has the highest level of interaction through the HYPERCON program were surveyed. These groups include:
- ACBM/NIST Computer Modeling Participants
- Academic partners and users
- Aggregates Producers
- Chemical Admixtures Producers
- Departments of Transportation
- Ready-Mixed Producers
- Standard Development Organizations

These stakeholder groups are described in some detail in Section 2.5.

Select responses from these surveys were also used to review the activities and organization of VCCTL. Two additional surveys were directly aimed towards present and past members of the VCCTL consortium.

3.3 Case Studies

 Yin (1984) defines the case study research method as "an empirical inquiry that investigates a contemporary phenomenon within its real-life context; when the boundaries between phenomenon and context are not clearly evident; and in which multiple sources of evidence are used." There are six main steps involved in case study development: 1) determine and define the research questions; 2) select the cases and determine data gathering/analysis techniques; 3) prepare to collect the data; 4) collect data; 5) evaluate and analyze data; and 6) prepare a report. Case studies tend to be complex because they involve multiple data sources and produce large amounts of data for analysis. When the main source of information for case study comparisons is derived from secondary sources, not all independent parameters are necessarily identical between case study entities. Thus, the analyst is put in a position of subjectively assigning information to comparable classifications which are developed *a posteriori* to match the available information found in secondary reporting sources. In an attempt to avoid bias that arises from such a process, this research bases case study analysis on first-person data determined by surveys. Case studies in this analysis of the HYPERCON program were designed to directly compare VCCTL and other leading cement/concrete research consortia. *A priori* questions were developed for the survey tool and asked relatively uniformly across consortium groups in order to establish a framework for thorough and unbiased comparison between case studies. The qualitative elements embedded in much of the explanatory data surrounding the case studies require a narrative comparison between case study entities. This was facilitated by developing a detailed database for recording key information for each case study, as discussed in Section 5.2.

3.4 Econometric Modeling: Qualitative Response Models

Qualitative response models (QRMs) allow econometric modeling of the factors underlying the decision-making of individuals or enterprises. The variables used in such modeling are derived from attribute-coded survey responses in combination with data from other sources. The simplest QRM is the linear probability model. Maximum likelihood analysis is used to obtain estimates of the parameters, and generally marginal effects are computed in these models. This modeling technique requires variable specification and data measurement. Additionally, the value of QRMs is questionable when the number of observations is less than 30.

Use of QRM is theoretically suited for application to this economic study; however, there are limitations on current HYPERCON data availability, precluding its immediate implementation. HYPERCON addresses a wide spectrum of cement and concrete industry needs through its research. Each component HYPERCON technical area is not comparable along the same variable specifications due to differences in scope, aim, and evolution. Once HYPERCON is further integrated and QRM comparable data has been gathered, it will become realistic to specify a "theoretical" QRM.

3.5 Grounded Theory

The development of HYPERCON between FY01 and FY09 was tracked using the conceptual framework of *grounded theory*. Grounded theory (Glaser and Strauss, 1967) offers a generalized way to view the evolution of a research program. The purpose of grounded theory is to develop theory about phenomena of interest, such as concrete performance. The theory (e.g. concrete performance models) needs to be grounded in observation. In a grounded theory approach to strategic basic research, the research begins with raising generative questions that guide the research process, but are intended to be neither static nor confining in nature. As the research team begins to gather data, *core theoretical concept(s)* are identified. Tentative linkages are developed between the core theoretical concepts and the data. These first steps can take years to complete. Subsequent research activities engage researchers in verification and summary; the effort tends to evolve towards one core concept that is central. Eventually, the research yields *conceptually dense theory* as new observation leads to refinement of tentative linkages and revisions in existing theory. This stage mirrors the concept of applied research to some extent; it is at this point that the core concept has been identified and fleshed out in detail and is disseminated to other industry players. The result of grounded theory is an extensively well-considered explanation for some phenomenon of interest, such as concrete performance.

Figure 3 shows HYPERCON's strategic basic research at an early stage of grounded theory development. The program objective is shown at the center: conceptually dense theory for concrete performance prediction. From FY01 to FY09, HYPERCON's strategic basic research has been concerned with identifying theoretical concepts for its five project components, gathering relevant data, and identifying linkages. Working inwards, HYPERCON in recent years has begun to consolidate through outputs linking these five components, such as the VCCTL, the ACBM/NIST Computer Modeling Workshop, and the Electronic Monograph. These outputs are beginning to form around a core theoretical concept for predicting performance of complex modern materials—Integrated Computational Materials Engineering (ICME). Indeed, even at this stage HYPERCON has become a leader for its computer modeling in the ICME community at large. The program is now poised to make a significant economic impact by strengthening and verifying its tentative linkages and further evolving its measurement science to a stage at which it becomes conceptually dense theory. Then HYPERCON's research outcomes will enable industry to implement concrete performance prediction in its integrated product development process, an outcome with potentially significant economic impacts.

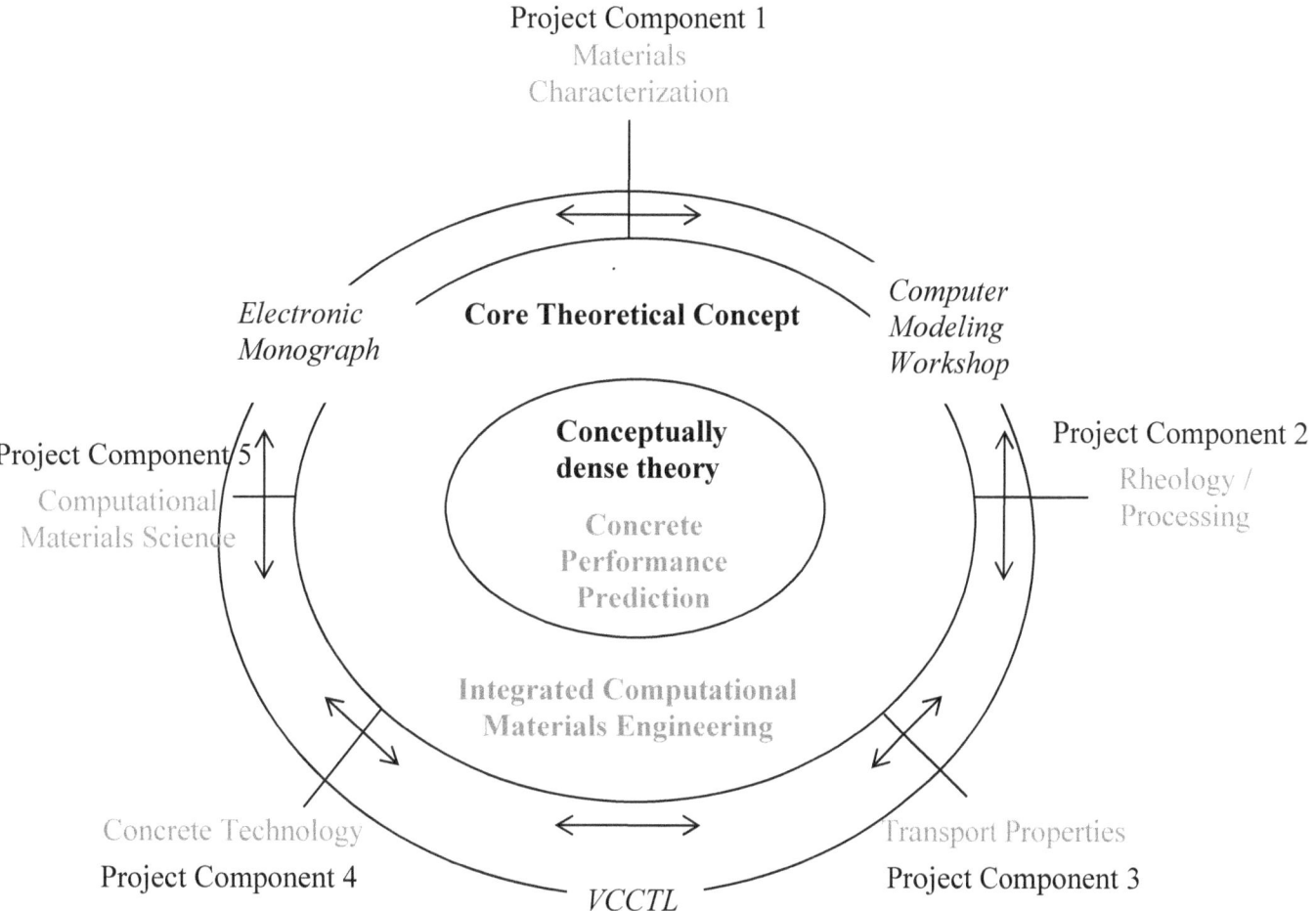

*Figure 3. HYPERCON in the Grounded Theory Context*

Grounded theory, which encourages continual re-working of linkages between project components and causal relationships, provides a good context for the establishment of the qualitative effects of HYPERCON research. The strategic basic measurement science from HYPERCON and the program history of its VCCTL consortium are both at a stage of development that permits qualitative impact assessment in this manner.

# 4.    HYPERCON Economic Analysis: Approach

The Economics of HYPERCON Project approach taken to this assessment is couched in *grounded theory* and draws primarily from data gathered through the use of surveys and case studies. This provides a rigorous framework using qualitative indicators to signify past program successes as well as to inform future HYPERCON project directions. This section provides details concerning the application of this framework to the HYPERCON program. Section 4.1 covers quantitative success metrics that are useful in this context. Qualitative metrics accessible through the survey process and use of case studies are discussed in Section 4.2. Data and results for both quantitative and qualitative metrics are presented and discussed in Section 5.

## 4.1 HYPERCON Quantitative Metrics

Thus far, the need for qualitative data and approaches to reviewing retrospective impacts of strategic basic research programs, like HYPERCON, has been stressed. Yet, there are quantitative data sets available that may be used as indicators of program success and that can provide guidance as to the most relevant questions to explore in a qualitative manner. Selected quantifiable metrics applicable to the HYPERCON program are identified and discussed below. Some of these metrics accrue to specific component projects of HYPERCON very clearly, while others apply to two or more component projects or address HYPERCON in general.

### Standard Reference Materials
NIST certifies and provides over 1,300 Standard Reference Materials (SRMs), which are used to perform instrument calibrations, to verify the accuracy of measurements, and to support development of new measurement techniques. SRMs are purchased by industry, government, and academic institutions, both nationally and internationally to advance research and development as well as to aid commerce activities.

From FY01 to FY08 several SRMs were developed as direct outputs of HYPERCON research. In particular, SRMs 2686, 2687, and 2688 address characterization for cement clinkers.[14] SRMs 46h and 114q (previously 114p) are used to calibrate fineness testing equipment according to ASTM Standard Methods and also present partial size distribution information. The SRM 114 series is an all-time best-seller for the NIST SRM program. The evolution of sales of SRMs directly enabled by HYPERCON research, by stakeholder group, geographic region, and year, can be traced.

### Guest Researchers / Academic Partnerships
HYPERCON technical areas have attracted guest researchers and graduate students from national and international institutions. At various points of development, some elements of HYPERCON projects have been researched in tandem with professors at leading U.S. institutions. A notable example is the partnership between NIST and Purdue University concerning the technical area of concrete technology. The number of guest researchers and academic institutional affiliations provide a quantifiable indicator of program success. However,

---

[14] Clinker is the solid material produced by the cement kiln stage of concrete production. Clinker is produced as lumps or nodules of 3 to 75 mm in diameter which will be ground to make Portland cement.

to understand the nature of these relationships and the extent of partnership and knowledge transfer to academic institutions from HYPERCON research, additional information of a qualitative nature is required.

### SDO Membership

Membership and clear HYPERCON contributions to SDOs interested in concrete and cement technology and standards is a partial indicator of retrospective HYPERCON research impacts and outcomes. Examples of SDOs in which HYPERCON research staff are active are ASTMInternational and ACI.

### Monograph usage

As one of the premier collections of strategic basic research on concrete available, *An Electronic Monograph: Modeling and Measuring the Structure and Properties of Cement-Based Materials*, holds a unique place within the academic, industrial, and governmental cement and concrete research communities. The Monograph was last revised on October 3, 2008 and includes approximately 4189 pages of printed text. Records of the approximate usage of Monograph are available for the economic study period on a monthly basis. The "usage" of Monograph is defined as the number of different computers that have accessed the Monograph monthly. The number of countries accessing the Monograph is recorded as well, although frequency is not a part of this metric.

### ACBM/NIST Computer Modeling Workshop participation

Statistics concerning the attendance changes from year to year for the ACBM/NIST Computer Modeling workshop are readily available. The data can be divided along primary sources of interest (e.g. faculty, industry, government, or student) and geographical region.

### Other Agency (OA) and Scientific and Technical Research Services (STRS) Funding

OA funds provide an important form of leverage for STRS allocated funding. OA funding levels, especially as a ratio to STRS for a project area, provide an indicator of the value that the outside funding enterprise puts on the research they are helping to fund. One example of this dynamic is VCCTL consortium member fees to fund and help direct the research that is funneled into the VCCTL software tool.

### VCCTL Consortium Membership

The VCCTL consortium began in FY01 and continues to draw interest from major industry stakeholders. Annual membership costs $40,000 per year. Representatives of VCCTL members attend bi-annual meetings and help to direct the elements of VCCTL and some of HYPERCON program research. Changes in membership (new member, continuing member, or end of membership) are one way to trace VCCTL consortium participation for each year during the study period. Yet, qualitative data developed through a survey provides greater context for understanding changes in VCCTL consortium membership.

### H-Factors

The H-Factor was proposed by Hirsch (2005) as a metric to qualify the impact and quantity of an individual scientists' research output. The index is based on the distribution of citations by a given researcher's publications: *a scientist has index h if h of his N papers have at least h citations each, and the other (N − h) papers have at most h citations each.* HYPERCON researchers give invited talks to technical and industrial groups, regularly attend conferences and workshops that are important to key stakeholders, and make regular contributions to refereed journals that are widely read within the cement and concrete industry. The majority of HYPERCON project leaders have been with the program from FY01 to FY09. Thus, H-Factors are a reasonable proxy for HYPERCON published research.

## 4.2 HYPERCON Qualitative Metrics

Qualitative metrics will provide the framework for linking quantitative success indicators to an understanding of the paths through which HYPERCON has impacted stakeholders in the concrete industry. This subsection describes the HYPERCON survey and case study methods and provides an overview of potential qualitative impact metrics.

The Paperwork Reduction Act (PRA) of 1995 requires that OMB approve each collection of information by a Federal agency before it can be implemented. This statute defines "collection of information" to be any identical questions posed to 10 or more members of the public, whether written, electronic, or oral. The survey methodology adopted by the Economics of HYPERCON project is in compliance with the PRA. The range of stakeholders involved in the various HYPERCON technical areas is such that an identical set of survey questions was not appropriate for more than nine individuals. For instance, the questions relevant to industrial users of VCCTL are fundamentally different from those which are appropriate to pose to potential academic users of the software.

## 4.2.1 HYPERCON Stakeholder Surveys

Each survey was developed to address a given Stakeholder—HYPERCON technical area pair, as defined in Figure 2 in Section 2.4. Key contacts for each Stakeholder group were identified by HYPERCON researchers. It is presupposed that the individuals contacted had some level of previous knowledge of HYPERCON research outputs, and in some cases will have partnered in some of the underlying research efforts. The overriding goal of these surveys was to determine to what extent HYPERCON impacts and outputs, by technical area, have affected stakeholders' interests in the cement/concrete industry.

Qualitative issues are difficult to assess on a strict year-to-year basis because they depend much more on human cogitative memory, which is not necessarily broken down annually. The back-drop of quantitative metrics determined on a yearly basis, however, should aid in developing annual qualitative estimates. Another timing-related issue to keep in mind is the considerable time lags expected from the year of a quantitative achievement (e.g. ASTM standards) to the time period in which it becomes important to a certain Stakeholder.

Economists conducting this study worked with HYPERCON researchers to determine the best phrasing for questions to ask each Stakeholder—HYPERCON technical area pair. Below are examples of qualitative issues that were translated into survey questions, often through links with quantitative metrics:

- The perceived value of partnership work with NIST.
- How HYPERCON strategic basic research has been used by the Stakeholder.
- What research activities the Stakeholder has undertaken to extend HYPERCON strategic basic research to applied research.
- The Stakeholder's rating of technical areas of HYPERCON research that are not directly related to their main HYPERCON technical area of interest.
- Estimated use value of SRMs that have been purchased.
- Perceived overall value of HYPERCON research applications and partnerships.

The NIST Program Office has documented the potential value of obtaining metrics based solely on *perceptual* measures of success in order to analyze the benefits of R&D activities (Dyer et al., 2006). The subjective assessment (via Likert ratings and explanations) of a given stakeholder of the *overall value* of HYPERCON research may provide context for the impact pattern revealed by other survey responses. For instance, if a respondent describes HYPERCON as having marginal overall value, but other responses reveal that the stakeholder actually uses HYPERCON research quite significantly, there may be an issue with branding and image, not the value of the HYPERCON research itself. There is a cyclical relationship between perceived value and actual usage of a product or R&D project research outcome. Thus, it is ideal to explore these relationships between stakeholders and HYPERCON research, especially in the context of informing future HYPERCON research dissemination modes.

While this study is primarily retrospective in nature, its results have been analyzed to some extent in support of future programmatic decision-making (Section 6.2). The idea is to help HYPERCON best address concrete performance prediction and measurement science needs of the U.S. industry into the future. To this point, during the survey process, stakeholders were asked to discuss what they see as the main strategic basic research need within their industry through open-ended questions. The data obtained through this genre of question is guided by modification of the Heilmeier questions:

- What is the most important strategic basic research need of your stakeholder group in the next 3-5 years? (coded as the attribute "mission or vision")
- How is this need addressed today? And in your opinion, what are the limits of current practice? (coded as "Market Research")
- In what way do you suggest NIST's HYPERCON program could help to solve this problem? (coded as "Intellectual capital")
- What difference do you think a successful outcome would make in 3-5 years time? (coded as "Value Added")
- What risks and roadblocks exist in your opinion? (coded as "Risk")
- How would success be measured? (coded as "Metrics")

4.2.2 VCCTL Surveys and Consortia Comparisons

The nature of the VCCTL research consortium and other associated VCCTL activities are differentiated from other HYPERCON component projects due to the structured partnerships and clear focus on bridging from strategic basic research to applied research activities. There are three other major concrete consortia with which to compare VCCTL: Summa, ACBM, and Nanocem.

SUMMA most closely mirrors the ICME efforts explicit in VCCTL. The Center for Advanced Cement-Based Materials (ACBM) has done joint work with HYPERCON and VCCTL in the past, one specific example being the yearly Computer Modeling workshop held at NIST. ACBM tends to be more focused on education and other programs throughout North America. NanoCEM is a European consortium, which connects industry and research institutions throughout the EU.

The case study presented in Section 5.2.3 describes the process by which research decisions are undertaken in each consortium and relevant linkages or overlaps with VCCTL research activities. Additionally, the differences and relative merits of each consortium by members who are active in more than one are explained. The case study comparisons are primarily based on research findings and detailed personal interviews with leading members of each consortium.

VCCTL consortium members have informally reported cost savings in the past from use of the VCCTL software tool. A major cement company reported savings of $ 1.2M in 2003 dollars over a three-week period enabled by the use of VCCTL software to determine the influence of "cement composition on mortar strength" and "fineness on mortar strength." A second major cement company reported a $ 750K savings in 2004 dollars directly attributable to virtual testing via VCCTL software. These numbers appear to provide a convincing case for the economic efficiency of this ICME approach, but the underlying assumptions and parameters of these anecdotal reports are typically not publicly available in traceable documentation. Companies are somewhat skeptical about sharing *how* they are using the VCCTL software tools because they think it may compromise their own intellectual property in some way. Thus, it is most effective to look at qualitative response metrics.

As previously mentioned, a survey tool aimed at current and past VCCTL members provides the basis for the case study comparisons between VCCTL and the other three consortia. Semi-structured surveys via telephone also garnered qualitative data necessary to address some of the major impact "channels" of VCCTL. The following are generalized impact metrics obtained through consultation with VCCTL consortium members, past and present:
- Level of industry support for VCCTL work (monetary and other support)
- Extent of use of the VCCTL software by consortium members
- Consortium members' strength of preference for each of the various technical areas VCCTL has explored from FY01 through FY09
- Percentage of VCCTL consortium members that rejoin / drop out of VCCTL each year
- The reasons motivating past members to drop out
- VCCTL consortium members' assessment of the balance of VCCTL research, between strategic basic research and applied research
- Relative success of various modes of transfer of VCCTL results (e.g. licensing, patents, and joint ventures)

As with the surveys covering non-VCCTL technical areas, it is useful to consider the needs of stakeholders in the near-term future to help inform VCCTL consortium directions. To this point, questions were included about near-term goals in the surveys specific to current VCCTL consortium members. Additionally, the HYPERCON program is scheduled to start an educational version of VCCTL, e-VCCTL, in FY10. On surveys sent to representatives of the Academic stakeholder group, their desires and specifications for the most useful e-VCCTL content and presentation was sought.

# 5. HYPERCON Economic Analysis: Data and Results

The scope of this report makes both analysis and reporting of data and findings in a consistent manner somewhat challenging. This section reports data related to quantitative indicators, supplemented by response patterns derived from qualitative methods. This process of data interpretation was enabled through the use of grounded theory. In most cases, findings were refined and specified based on answers to multiple surveys. There was a great focus on using Likert coding and other tools to make responses to similar questions in varying survey tools comparable. Subsequently, this process allowed for a wider base for comparison and when appropriate, a more statistically significant sample size from which to draw relevant statistics.[15]

Section 5.1 presents data related to the quantitative metrics identified in Section 4.1. Section 5.2 provides a rigorous discussion of generalized response patterns observed across survey answers and presents case study comparisons, following from the information provided in Section 4.2.

## 5.1 HYPERCON Quantitative Metrics

### 5.1.1 Standard Reference Material Usage

SRM sales were tabulated from FY02 through FY08 based on records accessed from the Measurement Services Division at NIST.[16] Those SRMs developed through HYPERCON research, as specified in Section 4.1, were evaluated yearly by unit and gross sales.[17] In the productivity literature it is implicitly mentioned that quantity indices over time are the desired "form" of output in productivity analysis, opposed to dollar amounts (Edridge, 1999). In this study we consider both. Costs of SRMs fluctuate over time due to inflation and other factors. These variations in unit price have been factored into the calculation of gross sales. SRM sales prices are set to comply with OMB Circular A-25, which requires full cost recovery.[18] The breakdown of SRM sales by SRM is provided in Table 1. Total sales of SRMs is just short of one million dollars, at $ 977,532. Sales of individual classes of SRMs from year to year are relatively constant. There is an exception for SRM 2688; sales reached a low point in FY04 at $ 7,462 and in FY05 grew to $ 21,139, its highest historical level.

Sales of all SRMs developed through HYPERCON research are aggregated by fiscal year in Table 2 (by unit sales) and Table 3 (by dollar sales). The majority of SRM sales is provided through the 114 (p and q) series, as the all-time bestsellers for the whole of NIST. During the period under review, sales of 114p and 114q totaled $ 687,195. Sales have increased steadily throughout the study period; the minimum sales is $ 101,729 (FY04) and the maximum is $ 194,687 (FY07).

---

[15] Due to OMB regulations on survey distribution, fewer than 10 responses were permissible for each survey.

[16] Raw data is available at: http://msd-i.nist.gov/srmreport/index.jsp. At the time of preparation of this report, FY09 data was very limited.

[17] SRMs 2686, 2687, and 2688 address characterization and particle size analysis for cement clinkers. SRMs 46h and 114q (previously 114p) are used to calibrate fineness testing equipment according to ASTM Standard Methods.

[18] Further information concerning the manner by which SRM sales prices are set is available in subchapter 5.19 of the NIST Administrative Manual: http://www-i.nist.gov/admin/mo/adman/519.htm#5.19.07.

While it is not possible to discern to which stakeholder group each of the buyers belongs, it is possible to distinguish foreign from domestic SRM sales. The percentage of overall sales to foreign addresses (proxy for foreign companies) has fluctuated between a high of 57 % (in FY02) to a low of 44 % (in FY07).

Questions specifically focused on the use of SRMs were asked in the surveys addressing stakeholders in the DOT and SDO communities. Of those responding that they indeed utilize SRMs in their research, 100 % have used a series 114 at some point during the research period. Fifty percent of these respondents had used each of the following: 46h, 2686, and 2688. One-third of respondents had used 2686a and 2687.

DOT and SDO stakeholders also were asked to indicate how their level of use of SRMs in general had changed during the study period. The possible responses ranged from "decreased significantly" to "increased significantly." As shown in Figure 4, only 1/8 of respondents answered that SRM usage had decreased at all. One-half of respondents reported that their usage had stayed relatively consistent during the period. Finally, 3/8 of the respondents reported a significant increase in their use of SRMs over time. These findings indicate that SRMs are indeed a successful and significant output of the HYPERCON program. The measurement science challenges of concrete/cement research are such that SRMs address basic needs of industry and enable vital research and development. The results show that addressing characterization and particle size analysis as well as calibrating for fineness testing through SRM development has successfully met users' needs. Specifically, SRMs helps allow for ASTM standards to be met consistently throughout the industry.

5.1.2 Guest Researchers / Academic Partnerships

HYPERCON supports a rich array of partnerships in government, the private sector, and academic research institutions. VCCTL research in particular has historically been supported by private industry initiatives and partnerships, which are reviewed in section 5.1.7.

The HYPERCON research program attracts international post-docs as well as mid-career and senior researchers as guest researchers, both from institutes around the U.S. and internationally. A number of these guest partnerships based at NIST extend well beyond a year. In some cases researchers have appointments as guest researchers based at NIST over two separate timeframes. The number of guest researchers working with HYPERCON researchers at the NIST facility for the whole year or part of the year is listed in Table 4. In a given year the ratio of guest researchers associated with a U.S. institution to those associated with a foreign institute is roughly 1:3.

For years, research in all HYPERCON strategic basic research areas has been conducted in collaboration with academic institutions, both nationally and internationally. Many of these institutions send short-term guest researchers to work at NIST. HYPERCON researchers also benefit from the use of some of the equipment and fundamental research based at these institutions. These longstanding HYPERCON partners are listed in alphabetical order in Table 5.

*Table 1. Unit Sales of HYPERCON-Related SRM by SRM number: FY02-FY08*

| SRM Number | Fiscal Year | Domestic | Foreign | Combined |
|---|---|---|---|---|
| 46H | FY02 | 1 | | 1 |
| | FY03 | -- | 2 | 2 |
| | FY04 | -- | -- | -- |
| | FY05 | -- | -- | -- |
| | FY06 | -- | 2 | 2 |
| | FY07 | -- | -- | -- |
| | FY08 | 242 | 90 | 332 |
| | *TOTAL* | *243* | *94* | *337* |
| 114 P & 114 Q | FY02 | 360 | 312 | 672 |
| | FY03 | 335 | 303 | 638 |
| | FY04 | 286 | 295 | 581 |
| | FY05 | 335 | 326 | 661 |
| | FY06 | 328 | 356 | 684 |
| | FY07 | 457 | 551 | 1008 |
| | FY08 | 307 | 457 | 764 |
| | *TOTAL* | *2408* | *2600* | *5008* |
| 2686 | FY02 | 67 | 30 | 97 |
| | FY03 | 46 | 28 | 74 |
| | FY04 | 30 | 20 | 50 |
| | FY05 | 46 | 50 | 96 |
| | FY06 | 40 | 42 | 82 |
| | FY07 | 29 | 41 | 70 |
| | FY08 | -- | -- | -- |
| | *TOTAL* | *258* | *211* | *469* |
| 2687 | FY02 | 63 | 33 | 96 |
| | FY03 | 46 | 28 | 74 |
| | FY04 | 30 | 20 | 50 |
| | FY05 | 46 | 50 | 96 |
| | FY06 | 40 | 42 | 82 |
| | FY07 | 29 | 40 | 69 |
| | FY08 | -- | -- | -- |
| | *TOTAL* | *254* | *213* | *467* |
| 2688 | FY02 | 62 | 31 | 93 |
| | FY03 | 42 | 25 | 67 |
| | FY04 | 24 | 21 | 45 |
| | FY05 | 57 | 61 | 118 |
| | FY06 | 39 | 46 | 85 |
| | FY07 | 29 | 46 | 75 |
| | FY08 | -- | -- | -- |
| | *TOTAL* | *253* | *230* | *483* |

*Table 2. Unit Sales of HYPERCON-related SRMs: FY02-FY08*

| Year | Domestic | Foreign | TOTAL |
|------|----------|---------|-------|
| FY02 | 553 | 406 | *959* |
| FY03 | 469 | 386 | *855* |
| FY04 | 370 | 356 | *726* |
| FY05 | 484 | 487 | *971* |
| FY06 | 447 | 488 | *935* |
| FY07 | 544 | 678 | *1222* |
| FY08 | 549 | 547 | *1096* |

*Table 3. Dollar Sales of HYPERCON-related SRMs: FY02-FY08*

| Year | Domestic | Foreign | TOTAL ($) |
|------|----------|---------|-----------|
| FY02 | 68,441 | 49,877 | *118,318* |
| FY03 | 59,587 | 47,518 | *107,105* |
| FY04 | 53,465 | 48,264 | *101,729* |
| FY05 | 68,327 | 73,234 | *141,561* |
| FY06 | 65,030 | 71,642 | *136,672* |
| FY07 | 86,407 | 108,280 | *194,687* |
| FY08 | 86,701 | 90,759 | *177,460* |
|  |  |  | *977,532* |

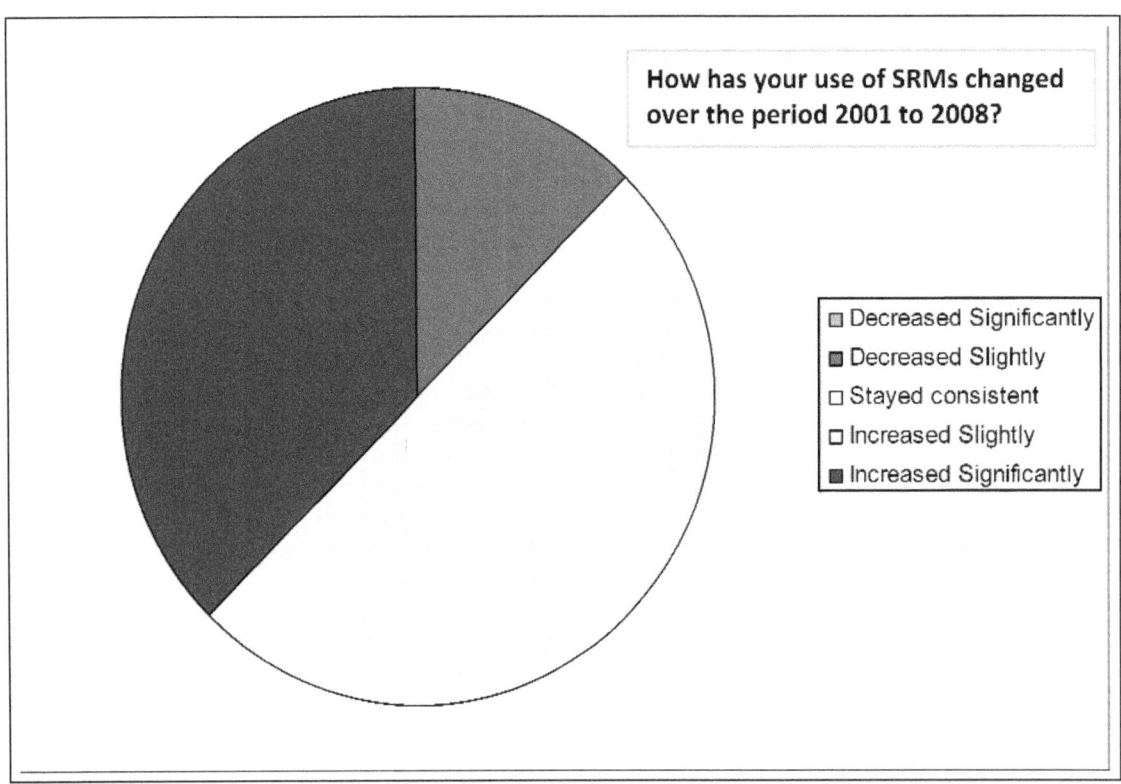

How has your use of SRMs changed over the period 2001 to 2008?

- Decreased Significantly
- Decreased Slightly
- Stayed consistent
- Increased Slightly
- Increased Significantly

*Figure 4. SDO and DOT Stakeholder Survey Responses: SRM Usage over Time*

*Table 4. HYPERCON Guest Researchers*

| Year | Number of Guest Researchers | Year | Number of Guest Researchers |
|------|------|------|------|
| 2001 | 7 | 2006 | 5 |
| 2002 | 9 | 2007 | 6 |
| 2003 | 5 | 2008 | 9 |
| 2004 | 5 | 2009 | 8 |
| 2005 | 5 | | |

*Table 5. HYPERCON Academic Collaborators*

| Domestic Partners | International Partners |
|------|------|
| Arizona State University | Delft University (Netherlands) |
| Cornell University | Denmark Technical University (Denmark) |
| Georgia Technical University | Laval University (Quebec City, Canada) |
| Purdue University | Middle East Technical University (Turkey) |
| Northwestern University | Queen's University of Belfast (United Kingdom) |
| Princeton University | University of Dijon (France) |
| Massachusetts Institute of Technology | University of Lausanne (Switzerland) |
| Northwestern University | University of New Brunswick (Canada) |
| Princeton University | University of Padua (Italy) |
| | |
| Tennessee Technical University | |
| Texas A&M | |
| University of Arkansas | |
| University of California, Berkeley | |
| University of Colorado, Boulder | |
| University of Florida, Gainesville | |
| | |
| University of Illinois, Urbana-Champaign | |
| University of Louisville | |
| | |
| University of Michigan | |
| University of Montana | |
| University of Texas, Austin | |
| Vanderbilt University | |
| Virginia Polytechnic Institute | |

The survey aimed at Academia provided further information concerning academic partnerships with HYPERCON. Eighty percent of those surveyed reported their affiliate intuition to be "active collaborative partners with [HYPERCON] at present or in the past." Among surveyed parties, the number of collaborations has grown throughout the study period. From FY02 through FY05, there were 4 active collaborators among those surveyed. The total number of collaborators remained constant, but there was minor turnover in the collaborating institutions. The number of active collaborations with HYPERCON grew to 80 % by FY07 and has remained constant. It is notable that collaborations with academic groups are generally multifaceted, spread over more than one HYPERCON focus area. The breakdown of academic collaborators is provided in Table 6 by HYPERCON focus area. Note that the sample size for the academic survey was too small to draw statistically significant conclusions from these results outside the fact that within single institutions, research collaboration foci tend to be highly differentiated.

*Table 6. HYPERCON Academic Collaborations by Technical Focus Area*

| HYPERCON Focus Area | Collaborations |
| --- | --- |
| Computational materials science | 8 |
| Concrete Technology | 1 |
| Durability and service life | 3 |
| Early-age cracking | 2 |
| Hydration | 2 |
| Materials Characterization | 5 |
| Rheology / Processing | 4 |
| Transport Properties | 2 |

It is also worth noting that some of the most meaningful collaborations are rather informal or significantly one-sided, with the NIST researcher providing expert advice to the academic institution. The following response is representative of many academic stakeholder responses:

> *"I have listed several areas of 'collaboration.' I know the whole group at NIST quite well, and often speak with them about research problems. My students and I occasionally visit, and even borrow equipment, but we do not have joint research proposals and we have not published papers together. We enjoy and benefit from close association with the NIST group, because they are leaders in the field."*

Due to the holistic nature of HYPERCON strategic basic research, there is value in asking academic stakeholders their overall impression concerning HYPERCON. The question was asked, "In your opinion, to what extent has NIST research met the research and education needs of the U.S. academic cement/concrete research community over the last eight years?" Given the generality of the question, the response choices were especially well specified: "More significant in the past, but has declined in recent years;" "Less significant in the past, but relevance has increased in recent years;" "Highly significant and maintained at a consistent level throughout the period;" "Somewhat significant and maintained at a consistent level throughout the period;" "Relatively insignificant and maintained at a consistent level throughout the period." Eighty percent of responses were categorized as "Highly significant and maintained at a consistent level throughout the period," while 20 % reported that the relevance was "More significant in the past, but has declined in recent years."

There have also been significant partnerships throughout time with institutions and individuals outside the academic community. These have generally been informal, but significant, exchanges between other consortia (as mentioned in Section 5.1.2) and with members of VCCTL (Section 5.1.7) One example is a European-based chemical admixture company that sent a representative to NIST for resident research for two months throughout the period FY02-FY09. Additionally, a NIST employee was sent to the company's site overseas for a month for onsite research. This joint work resulted in knowledge building and sharing for fundamental rheology studies, and is evidenced by the joint authorship of at least three well-received articles in the sub-field. This industry-NIST research partnership is a rarity regarding the ease with which it is traceable to an official and ongoing relationship between individual researchers on a shared research goal. Most such partnerships are rather ad hoc and informal, without direct, onsite research.

Further inquiry revealed that this industrial partner was financed by their company to come to NIST in person in order to expedite NIST research outcomes. As grounded theory suggests, other industrial partners were asked their opinion of the alacrity of NIST HYPERCON research outcomes. Essentially, the trend in responses is related to the positioning of the stakeholder relative to industry. Industrial responses from ready-mixed, chemical admixture, and aggregates companies reflected a respect for HYPERCON research, but a desire for greater speed in NIST applied research outcomes. Though, it should be noted that the structure of industry, requiring distinct applied research outcomes and allocating less than 1 % of its budget to applied basic research, is extremely different from academic settings and the structure of NIST. Academic partners with more official partnering of research activities tended to think that the speed of NIST HYPERCON research was fine, and held the research outcomes in high regard. Below is a paraphrased response to an open-ended response question in the Academic stakeholders survey.

> *"The research group under Dr. Garboczi is widely considered to be among the finest in the world. The staff scientists are recognized as world leaders in their respective fields and have made dramatic contributions, particularly in the area of computational modeling, to the field of concrete materials science. It is notable, however, that this group, while working on some of the most fundamental problems in the field, retains an application and field oriented and driven posture which bring their results to practice in a timely manner. The scientists are well integrated into relevant industrial organizations through various institutes, i.e. ACI, and their work is very visible."*

HYPERCON takes an active and systematic role in fostering extensions to the academic community. A main product in this outreach has been the Electronic Monograph. Additionally, the ACBM/NIST Computer Modeling Workshop attracts a number of PhD students and other players in academic cement/concrete research. The Electronic Monograph and the ACBM/NIST workshop are analyzed below.


5.1.3 Monograph Usage

As a premier collection of strategic basic research, the Electronic Monograph provides a log of strategic basic research for reference by the cement/concrete stakeholder communities. "Usage" is defined as the number of different computers (tracked by IP address) that have accessed Monograph in a given month. NIST computer records track all the "hits" to the Monograph and notes the country extension from each computer address; thus, the number of foreign countries accessing the Monograph can be tabulated by month.

Monograph access records are not complete for the entire study period. Due to hacker activity, March 2001 and April 2001 were not recorded and May 2001 provides only a partial count. Usage statistics for April 2003 and May 2003 also were not recorded due to hacker activity.[19] The majority of statistics for calendar years 2008 and 2009 are not available due to missed downloads from NIST computer records. To the extent possible, Monograph usage statistics are presented graphically in Figure 5. A trend line has been fit to the data. The linear average usage is calculated (after withholding months for which data is not available) to be about 10,300 hits per month. The usage of Monograph has generally increased since its inception. The historical minimum usage is 5749 in December 2001 and the maximum is 16,839 in July 2007.

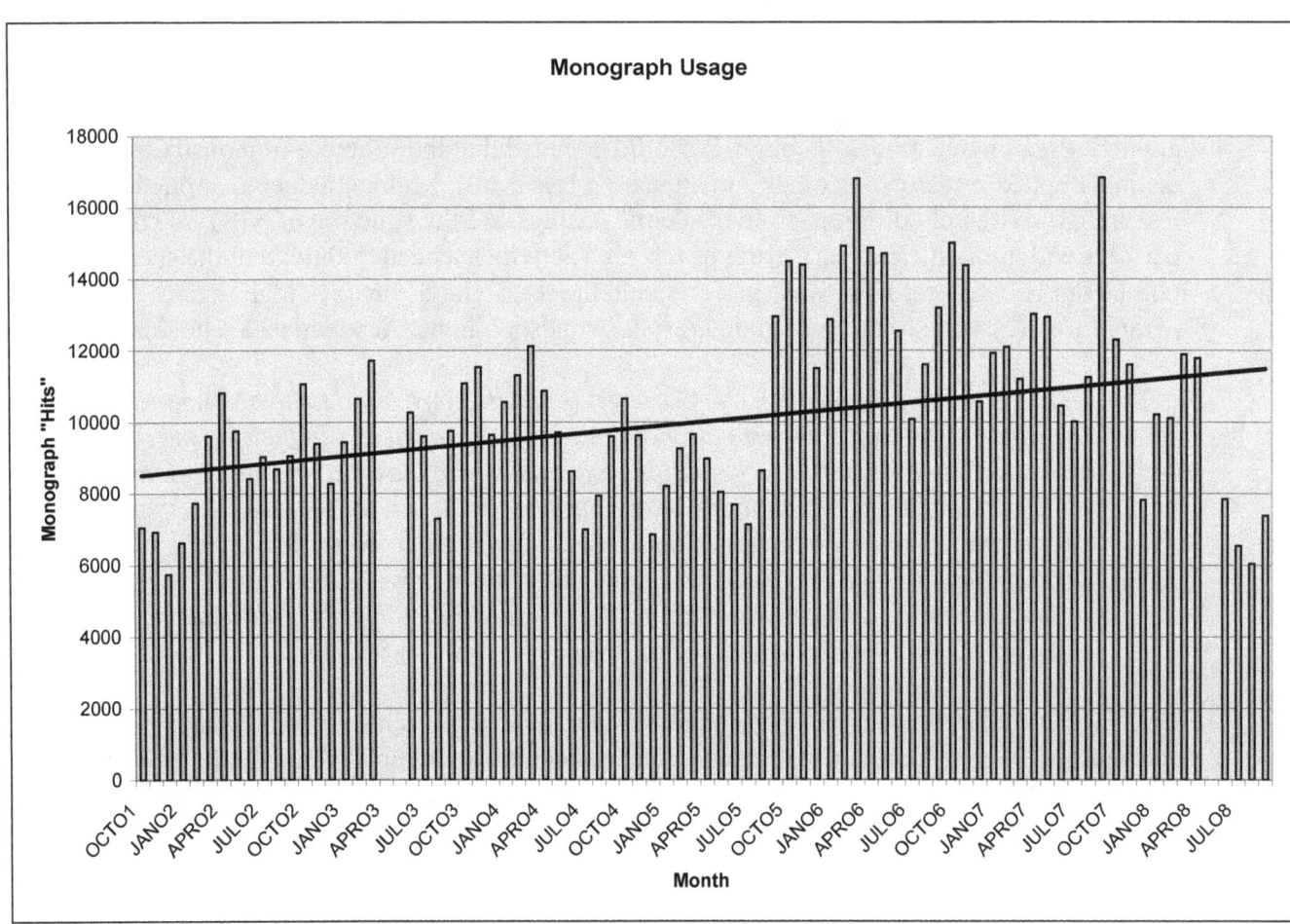

*Figure 5. Electronic Monograph Usage: FY02-FY08*

As an inclusive record of HYPERCON research findings, the Electronic Monograph is a relevant tool that cuts across stakeholder groups' interests. To this point, the following question appeared on multiple surveys: "In your opinion, how much has information learned through the NIST Electronic Monograph: Modeling and Measuring the Structure and Properties of Cement-Based Materials (http://concrete.nist.gov/monograph) affected your work over the last eight years?"

---

[19] Approximate usage levels could not able to be extrapolated for missing months due to high levels of variation from month to month.

The response possibilities were: "significant impact," "somewhat significant impact," "minor impact," "no noticeable impact," and "no use." The results are totaled for respondents from the following surveys: Chemical Admixtures, Aggregates, ACBM/NIST Modeling Workshop, and Academia. Just over 40 % of all respondents believe that Monograph has had "significant impact," while 30 % report "somewhat significant impact." Ten percent of those surveyed had not used Monograph. The aggregated responses are shown in Figure 6. The responses were reviewed for each separate stakeholder group. From industry (chemical admixtures and aggregates), 20 % found Monograph to have "significant impact," while the level rose to 55 % for the Academic community (Academia and ACBM/NIST Workshop).

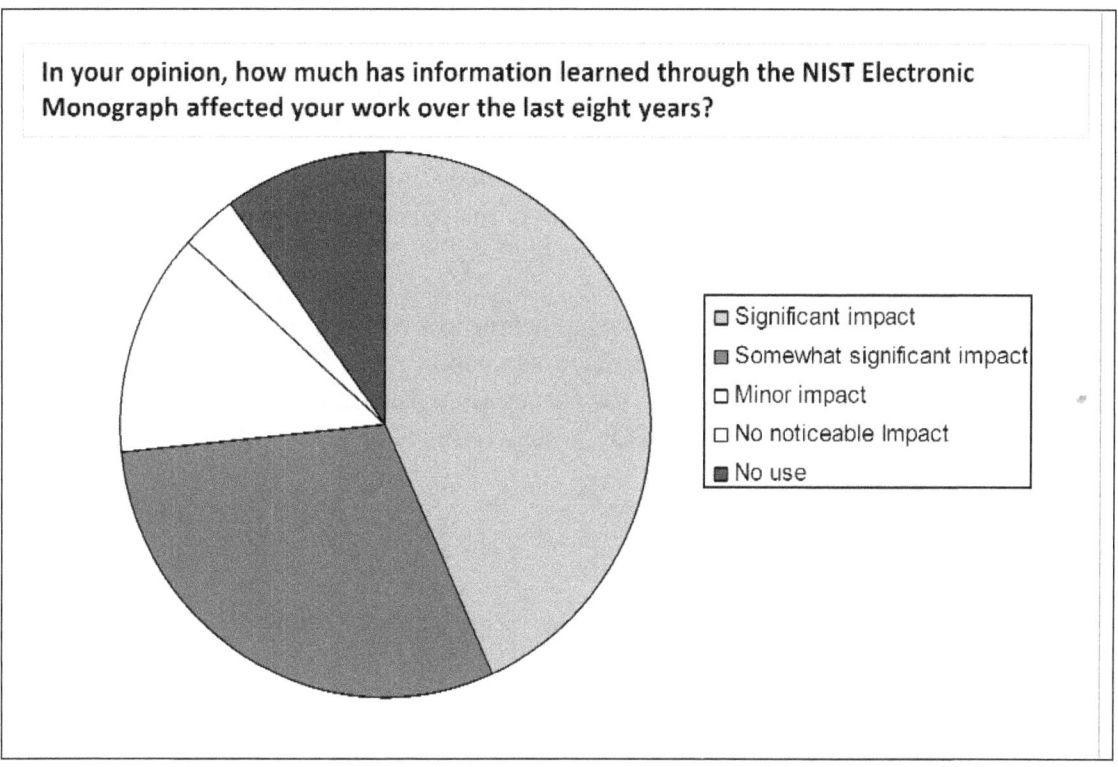

*Figure 6. Electronic Monograph Survey Response: Significance of Impact*

5.1.4 ACBM/NIST Computer Modeling Workshop participation

The ACBM/NIST Computer Modeling Workshop was historically a joint product of NIST HYPERCON research and the ACBM consortium. The Workshop has since 2000, been administered and organized entirely by the HYPERCON research team.

For the 209 workshops to date, the total attendance was 5774 individuals, representing 23 countries.[20] During the FY01-FY09 period, 290 individuals took part in the annual workshops[21] (2/3 from U.S. institutions and 1/3 from foreign institutions).

---

[20] Australia, Belgium, Brazil, Canada, Chile, Czech Republic, Denmark, France, Germany, Italy, Japan, Korea, Mexico, Netherlands, Norway, Poland, Portugal, Singapore, South Africa, Spain, Sweden, Switzerland, United Kingdom.

These attendees are distributed among academia, industry, and government as follows:
- Doctoral and Post-doctoral students: 153
- University Faculty: 48
- Industrial researchers: 60
- U.S. Government researchers: 29

In our survey aimed towards participants of the ACBM/NIST Computer Modeling Workshop, the distribution of respondents was roughly proportional to this reported distribution. Generally, about 2/5 of workshop participants are specifically interested in computer modeling techniques, with no direct relationship to cement or concrete research. This fact provides a challenge as to how best to present techniques and the extent of the discussion that should be generalized away from cement/concrete. Historically the workshop lectures present HYPERCON research topics in the cement/concrete context, while allowing some abstraction to other applications. For example, a popular lecture, "Analyzing Particle Shape in 3-D," has been lauded by participants for the universality of its topic to applications in a number of scientific fields interested in the increased use of ICME techniques.

Conducted over three days, the annual workshop format allows for a mixture of longer tutorial sessions and short technical talks by the participants in addition to the core lectures by HYPERCON researchers. Respondents to the survey were asked to indicate their areas of interest with regards to the main HYPERCON research topics presented during the Workshop. This breakdown is given in Table 7.

*Table 7. ACBM/NIST Modeling Workshop Survey: Workshop Topics of Interest*

| Workshop Topic | Response Count |
|---|---|
| Materials Characterization | 7 |
| Rheology / Processing | 3 |
| Transport Properties | 6 |
| Concrete Technology | 3 |
| Computational material science | 5 |
| Early-age cracking | 3 |
| Durability and service life | 7 |
| Heat Transfer | 1 |
| Novel Concrete Uses | 1 |

Almost 80 % of respondents had participated in the Workshop over multiple years (2, 3, or 4 workshops). These respondents were pleased with the update of presented information from year to year. When asked, "In your opinion, how much has information learned through the annual ACBM/NIST Computer Modeling Workshop affected your work over the last eight years?," 5/9 responded that there was at least a "somewhat significant impact." Four-ninths of respondents reported "minor impact," while there was no report of "no noticeable impact."

---

[21] This is based on hard data for 2001-2008 on the actual attendees, and on the preliminary attendance list for 2009.

A number of participants report that the Workshop has introduced them to a number of other HYPERCON-developed tools, such as the Monograph. Four-ninths of respondents state that they have developed research collaborations through interaction at the Workshop. Of these, 75 % have been with NIST researchers exclusively. Additionally, for a number of participants, the Workshop serves as a first introduction into the use of other HYPERCON-developed tools, such as the Monograph. For those reporting first use of such tools after the Workshop, use has been sustained through the present.

All surveyed parties were asked to provide critiques of the Workshop through an open-ended response. There were a number of suggestions:
- Provision of an explicit tutorial-level display of the developed software/technologies.
- Additional chemistry review portion before detailed discussion sections.
- Addition of problem solving sessions to test knowledge and basic understanding.
- Hands-on modeling opportunities throughout the Workshop based on actual computer interface usage.
- Sessions exclusively designed for specific stakeholder groups, e.g. transportation agency personnel.

A guest researcher in a computing group at NIST joined the 2008 Workshop and had some interesting feedback about its structure. The overriding theme was needs for improvement in the balance between theory background and the actual use of the HYPERCON-developed software tools. This researcher had worked on re-scripting sections of the software, so had an advantage in understanding elements that were not explained. His feeling was that participants from the cement/concrete research field already knew much of the technical information presented, but could have benefitted from more emphasis on software use. Other participants were there to explore ICME techniques and computer modeling/simulation in a general sense, so also did not require such technical detail relating to cement/concrete. To this point, three of the surveyed individuals did suggest that the Workshop participants be divided at the beginning or at least at certain points based on their background knowledge and their goals for Workshop participation.

In general, there was very positive feedback about the Workshop over the past eight years. Some examples follow:

> I've enjoyed the past conferences, and I have found the NIST representatives to be both very knowledgeable and helpful.

> Ed Garboczi's group at NIST is and has been a very valuable resource and made significant contributions to the materials science of concrete. The NIST team is very easy to work with and always friendly, professional, and responsive to requests during the yearly conferences. I consider Ed Garboczi's group to be the central resource for cement/concrete research in the United States. His technical staff is involved or aware of the current national and international work in the field and they are knowledgeable of important historical work in the field.

## 5.1.5 SDO Membership

Most HYPERCON project researchers take an active role in at least one of the main SDOs related to cement/concrete research activities (ASTM International and ACI). There is a high level of commitment to leadership on ASTM Committee C01 on Cement as well as ASTM Committee C09 on Concrete and Concrete Aggregates. HYPERCON researchers have held Chairs on ACI committees, and have been a member of the ACI Technical Advisory Committee (TAC). There is a strong history of standards based on HYPERCON research being readily adopted by these SDOs.

Due to the structure of some SDOs and the research activities that are actually relevant to various groups, some HYPERCON research areas have been more applicable to standards than others. This is a caveat of direct comparison of SDO membership between given HYPERCON researchers. To which point, we look only at the aggregate membership of the HYPERCON group in ACI and ASTM committees. The division of membership is given by individual as an organizational tool, but individuals are not identified. SDO committee participation by HYPERCON researchers is provided in Table 8.

*Table 8. SDO Committee Participation by HYPERCON Researchers*

| ASTM | ACI |
|---|---|
| C01 Frost Resistance sub-committee | 201 Durability |
| C01.10 Hydraulic Cement | 225 Cements |
| C01.22 Workability | 231 Early Age |
| C01.23 Compositional Analysis – Chair | 236 Materials Science |
| C01.25 Fineness | 236 Materials Science |
| C01.26 Heat Hydration | 236 Materials Science |
| C01.28 Sulfate Content | 237 Self-consolidating |
| C01.29 Sulfate Attack – Vice-Chair | 238 Workability – Chair |
| C01.31 Volume Change | 308 Curing |
| C01.48 Admixture Interaction | 309 Consolidation |
| C01.90 Executive subcommittee | 349 Nuclear structures |
| C01.92 Administrative Coordination | 552 Grouting |
| C01.95 Coordination of Standards | 201 Durability |
| C09 Concrete and Concrete Aggregates | 225 Cements |
| C09.21 Lightweight Aggregates | 231 Early Age |
| C09.65 Petrography | 236 Materials Science |
| C09.66 Concrete's Resistance to Fluid Penetration | |
| C09.67 Resistance to the Environment | |
| C09.68 Volume Change | |

The number of standards activities related to HYPERCON research is impressive; however, the relative value placed on NIST's role in these standards within and outside of the standard-setting community is a pivotal indicator of success for this metric. This information can be gleaned from response patterns observed in the survey aimed towards SDO stakeholders.

A number of Likert-scale response questions were aimed at HYPERCON project areas that generally have been recognized as providing requisite research for a number of standards regarding cement/concrete. The response possibilities for these questions were set at: "highly applicable;" "somewhat applicable;" "not directly applicable;" "not directly applicable, but of significant future interest;" and "not much foreseeable future applicability or interest."

Some of the survey questions asked respondents about specific project areas under the HYPERCON program. When asked to indicate the "extent to which NIST's work on increased service life and durability has been applicable to or informed your work on standards in the industry," all respondents chose "somewhat applicable." Respondents were also asked the "extent NIST's work on rheology/processing using developed metrological methods for accurately measuring and modeling the rheology of cement paste, mortar and concrete" has informed standard-setting work. Twenty-five percent of respondents concluded "somewhat applicable," while 50 % chose the response "not directly applicable" and 25 % specified that the research has been "not directly applicable, but of significant future interest."

SDO stakeholders also were asked a number of questions about HYPERCON in a general sense. Overall, 25 % found that over the last eight years, NIST's research has been "highly relevant and maintained at a consistent level" with regards to the needs of the cement and concrete standards community. The balance of 75 % responded that NIST research has been "somewhat relevant and maintained at a consistent level." Additionally, respondents were asked whether "the standards to which NIST has been the prime contributor/motivator in the past 8 years have been important standards for the cement/concrete community?" Twenty-five percent responded that these standards have been "highly important" standards, while 75 % classified them as "of medium importance" on average.

Respondents indicated that in some ways standards have been a medium for knowledge transfer from strategic basic research into the cement/concrete industry. Recognizing that the needs for standards are highly varied throughout the cement/concrete industry, respondents were asked to indicate the extent to which HYPERCON research has met the technical needs in their specific area over the past eight years. Twenty-five percent of respondents answered that HYPERCON research had been "more significant in the past, but has ebbed off in recent years." Just over twelve percent responded that it had been "less significant in the past, but relevance has increased in recent years." The majority of 37.5 % found HYPERCON research to be "highly relevant and maintained at a consistent level throughout 2001-2008." Finally, the remaining 25 % of respondents reported that they found HYPERCON research "somewhat or less relevant, but maintained at a consistent level throughout 2001-2008."

HYPERCON research is focused in some areas that are relatively new (exploratory) to the established cement/concrete industry. Thus, these areas are a medium for technology transfer, but at the same time are still far from the formalized standard adoption process. There is a healthy combination of standards being set through research that is more mature in the market. For example, a HYPERCON researcher was awarded the ASTM P.H. Bates and the NIST Bronze Medal for work on developing a new ASTM Standard Test Method.

Leadership and past contributions to standard-making by HYPERCON research staff are not the only indicators of the value SDOs place on HYPERCON research. Another indicator is the degree to which SDO's rely on NIST scientific expertise and its impartial role in standards making. For example, there are a number of areas for which SDO stakeholders would like to see increased research from HYPERCON. Specifically, these strategic basic research areas were highlighted as likely having a place for standards designation in the next five years and were specified as needing NIST expertise to help address industry non-consensus. These are listed below as supplied by respondents in an open-ended query form:
- o Direct phase determination
- o Eliminating prescriptive requirements for concrete in favor of performance requirements (i.e. durability tests)
- o Optimum burning conditions in kilns
- o Effective use of supplementary cementitious materials
- o Modeling of durability
- o Improved characterization for performance prediction, especially in novel cements (e.g. those with lower clinker factors)
- o Effects of pre-hydration on cement performance
- o Categorization of significantly different hydration behavior
- o Cement characterization and phase composition analysis improvements
- o Participation in interlaboratory studies on standards

5.1.6 Other Agency and Scientific and Technical Research Services Funding

Other Agency (OA) funds provide an important form of leverage for internal NIST funding (known as STRS). OA funding levels, especially as a ratio to STRS for a project area, can provide an indicator of the value that the outside funding enterprise puts on the research. One example of this dynamic is VCCTL consortium member fees to fund and help direct the research that is funneled into the VCCTL software tool. The level of BFRL investment (STRS) and leveraged R&D funds (OA and VCCTL) is reported in Table 9. Please note that the values for 2009 are estimated.

Over the study period, there has been some variation in STRS funds, but they have remained relatively consistent. Due to the relative difference in magnitude between leveraged R&D and STRS means that any increase in OA or VCCTL funds is relatively significant. Increases in VCCTL funding is directly related to VCCTL membership levels from year to year, which is discussed in Section 5.1.7. Looking specifically at OA funds, there was significant growth in funds from 2007 to 2008, which is estimated to continue in the coming years. In the face of the current economic downturn, such growth is encouraging and validates the value of HYPERCON research.

*Table 9. STRS and Leveraged R&D Funds per year\**

| Year | STRS | $K Leveraged R&D | |
|---|---|---|---|
| | | OA | VCCTL |
| 2001 | 1118.0 | 171.3 | 83.0 |
| 2002 | 1391.3 | 101.5 | 147.4 |
| 2003 | 1688.7 | 69.4 | 148.1 |
| 2004 | 1517.8 | 262.1 | 507.4 |
| 2005 | 1496.2 | 245.9 | 196.4 |
| 2006 | 1212.5 | 232.0 | 632.8 |
| 2007 | 1613.9 | 146.5 | 371.5 |
| 2008 | 1588.7 | 457.6 | 368.1 |
| 2009 | 1548.7 | 500.0 | 416.9 |
| TOTAL | 13175.8 | 2186.3 | 2871.6 |

\* Note that these values are not discounted by year.


5.1.7 VCCTL Consortium Membership

The VCCTL consortium has always maintained at least 6 major industry stakeholders with paid memberships. At its largest membership there were nine industry members. Annual membership fees have been maintained at $40,000 per year since FY01. The level of fluctuation in membership by year is provided in the table below.

Current members of VCCTL (as of 3 February 2009) are shown in Table 10.

*Table 10. VCCTL Consortium Membership as of February 2009*

| Organization | Stakeholder Type | Membership Start Date | Country |
|---|---|---|---|
| BASF Admixtures, Inc. | Chemical Admixtures | February 1, 2004 | United States |
| FHWA Turner-Fairbanks Highway Research Center | U.S. Federal Highway Laboratory | June 30, 2008 | United States |
| Mapei | Chemical Admixtures | February 1, 2008 | Italy |
| Ready-Mixed Concrete Research and Education Foundation | Ready-Mixed Concrete | February 1, 2003 | United States |
| Sika Technology A.G. | Chemical Admixtures | February 1, 2003 | Switzerland |
| W.R. Grace & Company | Chemical Admixtures | February 1, 2001 | United States |

The majority of current VCCTL members fall under the chemical admixture stakeholder category. State DOTs have become much more interested in the prospect of membership. Yet, potential new members and a number of those with lapsed memberships have identified "tough economic conditions in the industry" as a main motivation to not take part in VCCTL at present. Variation in operating budgets among VCCTL stakeholders and other economic constraints brought on over the past two years make VCCTL membership an unreliable quantitative metric. The impressions of current and past VCCTL members are discussed in full in Section 5.2.12.

5.1.8 H-Factors

H-Factors are used as a metric to qualify the impact and quantity of and individual scientist's research output. H-Factors for HYPERCON researchers were determined through the use of "Web of Science[22]." In addition to H-Factors, the total number of citations and total number of papers were tabulated for each researcher, as well as the average citations per item. These results are reported in Table 11. The researchers are coded by R1 through R7.

*Table 11. H-Factors for HYPERCON and its Researchers*

| | Articles Found | Sum of Times Cited | Avg. Citations per Item | H-Factor |
|---|---|---|---|---|
| All researchers, 1970-2009 | 248 | 4,755 | 19.17 | 38 |
| All researchers, 2001-2009 | 110 | 1,072 | 9.75 | 18 |
| R1 | 9 | 31 | 3.44 | 4 |
| R2 | 31 | 353 | 17.26 | 12 |
| R3 | 41 | 293 | 7.15 | 10 |
| R4 | 39 | 398 | 10.21 | 10 |
| R5 | 10 | 72 | 7.20 | 4 |
| R6 | 20 | 93 | 4.65 | 6 |
| R7 | 17 | 164 | 9.65 | 9 |

As a point of reference the pertinent statistics are totaled in Figure 7 and Figure 8 for *all* journal articles including any HYPERCON researchers as authors since 1970. When limited to the 2001 to 2009 time period (Figure 9 and Figure 10), these statistics tend to be lower than those determined over an individual's entire career because of the constraints on publication and citation dates (papers have to be written and cited in the eight year timeframe). This report analyzes strategic basic research conducted and reported between 2001 and 2009; thus, the main analysis for each HYPERCON researcher was conducted for these dates.

The HYPERCON research group self-identified[23] the 8 most pivotal papers for the field authored by a member of the group in the past eight years. There is a not a good way to control for differences in citations numbers between years due to the cumulative progression from year to year within the 8 year span. These papers are listed, along with the number of citations attributed to each, below. It is important to note that the applicability of research in cement/concrete to that of different scientific fields is widely variable. Thus, the citations of some researchers may naturally be much higher due to wide applicability across fields.

---

[22] H-Factors were also calculated on Scopus and Google Scholar. It was found that "Web of Science" had the most complete coverage of journal publications.
[23] Those in the cement/concrete field, as opposed to having wider applicability to other fields.

| Publication Reference | Number of Citations |
|---|---|
| Bentz, D.P., Mizell, S., Satterfield, S., Devaney, J., George, W., Ketcham, P., Graham, J., Porterfield, J., Quenard, D., Vallee, F., Sallee, H., Boller, E., and Baruchel, J., The Visible Cement Data Set, *Journal of Research of the National Institute of Standards and Technology*, 107 (2), 137-148, 2002. | 24 |
| C. F. Ferraris, N. S. Martys, Relating Fresh Concrete Viscosity Measurements from Different Rheometers, J. Res. Natl. Inst. Stand. Technol. 108 (3 ), 229 -234 (2003) http://www.nist.gov/jres | 8 |
| K.A. Snyder and D.P. Bentz, Suspended hydration and loss of freezable water in cement pastes exposed to 90% relative humidity, Cement and Concrete Research 34(11), 20452056, 2004. | 17 |
| J.W. Bullard, "A Determination of Hydration Mechanisms for Tricalcium Silicate Using a Kinetic Cellular Automaton Model," J. Am. Ceram. Soc. 91 [7] 2088-2097 (2008). | 1 |
| J.W. Bullard, "A Determination of Hydration Mechanisms for Tricalcium Silicate Using a Kinetic Cellular Automaton Model," J. Am. Ceram. Soc. 91 [7] 2088-2097 (2008). | 1 |
| P. Stutzman, "Scanning Electron Microscopy Imaging of Hydraulic Cement Microstructure," Cement and Concrete Composites, 26 (8), 957-966 (2004). | 14 |
| A.P. Roberts and E.J. Garboczi, "Computation of the linear elastic properties of random porous materials with a wide variety of microstructure," Proc. Roy. Soc. Lond. A 458, 1033-1054 (2002). | 36 |
| E.J. Garboczi, Three-dimensional mathematical analysis of particle shape using x-ray tomography and spherical harmonics: Application to aggregates used in concrete, Cem. Conc. Res. 32, 1621-1638 (2002). | 57 |
| N. S. Martys, "Study of a dissipative particle dynamics based approach for modeling suspensions," J. of Rheology 49, pp. 401-424, 2005. | 23 |

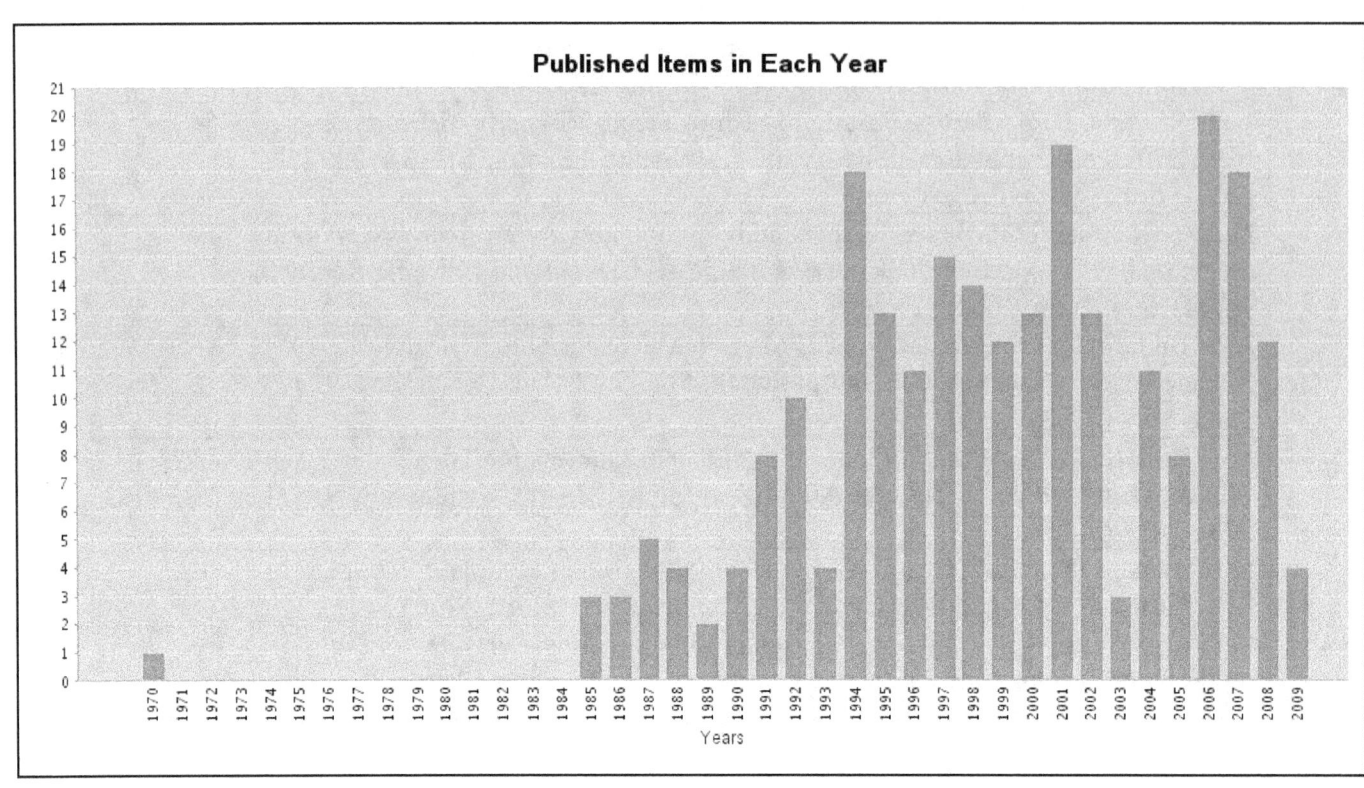

*Figure 7. HYPERCON Publications: 1970:2007*

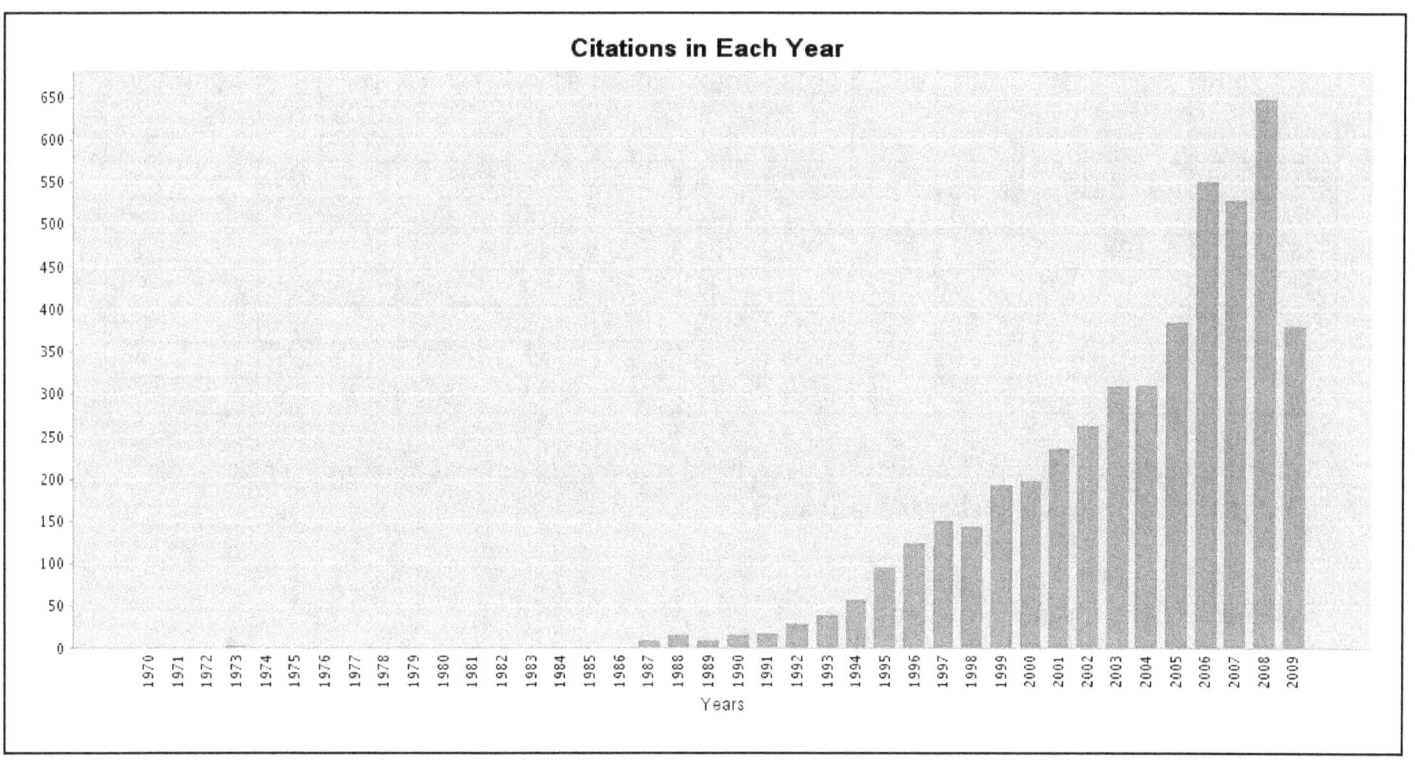

*Figure 8. Citations to HYPERCON Publications: 1970-2009*

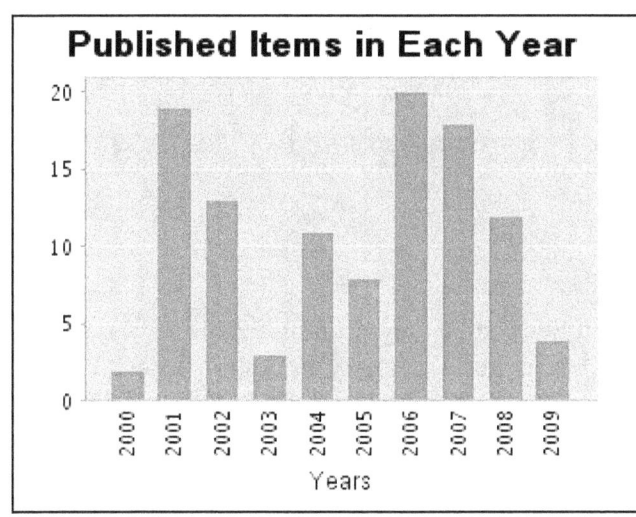

*Figure 9. HYPERCON Publications: 2000-2009*

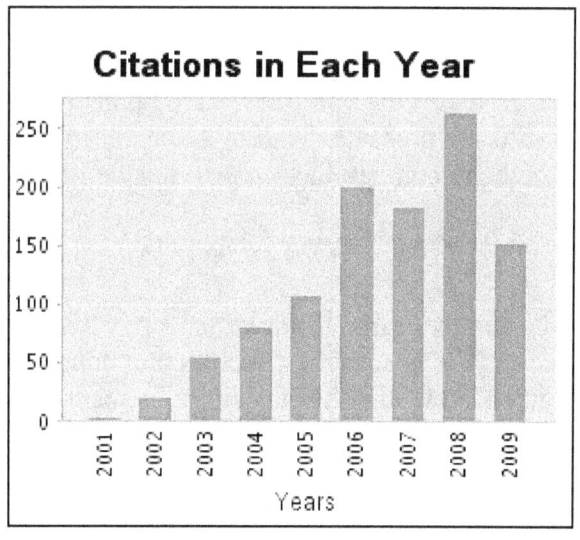

*Figure 10. Citations to HYPERCON Publications: 2001-2009*

49

*Table 12. Frequency of Citations to HYPERCON Research by Researcher: 2001-2008*

|  | 2001 | 2002 | 2003 | 2004 | 2005 | 2006 | 2007 | 2008 | 2009 | TOTAL | Total (2001-2008) | AVERAGE |
|---|---|---|---|---|---|---|---|---|---|---|---|---|
| R1 | 0 | 1 | 1 | 4 | 2 | 5 | 9 | 3 | 6 | 31 | 25 | 3 |
| R2 | 0 | 1 | 4 | 9 | 8 | 7 | 13 | 22 | 8 | 72 | 64 | 8 |
| R3 | 0 | 0 | 0 | 1 | 6 | 19 | 23 | 30 | 14 | 93 | 79 | 10 |
| R4 | 1 | 6 | 11 | 10 | 12 | 35 | 32 | 38 | 20 | 165 | 145 | 18 |
| R5 | 2 | 3 | 6 | 14 | 27 | 41 | 59 | 79 | 62 | 293 | 231 | 33 |
| R6 | 1 | 1 | 2 | 5 | 18 | 83 | 90 | 126 | 72 | 398 | 326 | 44 |
| R7 | 1 | 13 | 35 | 50 | 63 | 121 | 72 | 119 | 61 | 535 | 474 | 59 |

Note that based on citation frequency by year, there is a general upward trend in the quality of HYPERCON's research as shown in Table 12.

## 5.2 HYPERCON Qualitative Metrics

### 5.2.1 HYPERCON Stakeholder Surveys: Overview

In this subsection the results of each stakeholder survey are briefly discussed. The primary focus in this section is on narrative responses provided by the surveys. In most cases, these specify the areas of strength and weakness within the HYPERCON program related to the needs of the various stakeholders. Some stakeholders were asked to specify areas within their work which would benefit the most from new or increased HYPERCON research. In these cases, the responses are discussed in Section 6.2 as areas of future HYPERCON research that could benefit U.S. industry.

A number of the survey responses were discussed in the previous subsection because they directly support and provide clarification for some of the quantitative metrics. Here, the remainder of the questions based upon opinions and coded via a Likert-scale are reviewed. The results are presented both by stakeholder survey, and by question theme for those response areas which are comparable between stakeholders.

### 5.2.2.Academia Survey Responses

The survey geared towards Academia was covered in Section 5.1 since a number of HYPERCON activities, such as the Monograph and Computer Modeling Workshop are well-embedded within the curricula of cement/concrete research departments.

The importance of HYPERCON strategic basic research to Academia is evidenced by the fact that 100 % of respondents indicated that "NIST's cement/concrete research program as a whole has had a **_high impact_** on basic science knowledge in the field over the past eight years[24]." When asked "To what extent do you see NIST as a U.S. and worldwide leader in cement and concrete research," there was a 100 % response of "NIST is a world and US leader in cement and

---

[24] The response options included: "high impact," "moderate impact," "little impact," and "no direct impact."

concrete academic research[25]." However, the response pattern changed slightly when asked the "extent to which NIST research met the research and education needs of the U.S. academic cement/concrete research community in the past eight years." Thirty percent of respondents indicated that HYPERCON strategic basic research had been "more significant in the past, but has declined in recent years," while 70 % feel that it is "highly significant and maintained at a consistent level throughout 2001-2009." One respondent clarified this dichotomy in perceptions of HYPERCON's relevance to academia: It appears that recently there has been a minor, but perceivable shift in academic focus from the most basic research to include technology transfer.

Outside of collaboration, the majority of respondents specified an appreciation for the strong role that computer models and publicly available, Internet-accessible tools provided by NIST play in their curriculum formulation. 50 % of respondents are "currently using HYPERCON programs and tools in one or more classes;" 50 % are "not currently using these tools, but have used then in the past."

The level of importance HYPERCON plays in education, especially at the graduate level, is most readily understood through examination of the open-ended responses:
- o "The modeling tools are an invaluable resource for both education and research. The documentation is quite thorough and the models are comprehensive and usable."
- o "Excellent materials that can help to push students to look in new ways. I wish I had more time to use them. Though the new virtual site [educational VCCTL software] will be useful."
- o "The tools developed at NIST have been invaluable in the training of my graduate students. Recently, due to budget cutbacks and increasing enrollments, we've begun using the virtual testing lab in our undergraduate materials course. It has been exceptionally useful for teaching the complex relationships between materials selection and performance in cement-based materials."
- o "I said that I have not 'actively collaborated,' yet I listed several areas of "collaboration", because I'm not sure what sort of activities you intended to include. I know the whole group at NIST quite well, and often speak with them about research problems. My students and I occasionally visit, and even borrow equipment, but we do not have joint research proposals and we have not published papers together. We enjoy and benefit from close association with the NIST group, because they are leaders in the field."
- o "Dale Bentz has been especially supportive of my research group's particular needs for slight modifications to existing NIST software. His willingness to accommodate our requests has allowed us to make substantial contributions to the industry in the area of concrete bridge deck management."

5.2.3 Aggregates Survey Responses

Overall, the aggregates community appears to be pleased with the relevance of HYPERCON research to their industry needs. A number of survey questions regarded highly technical issues concerning particle shape characterization. As shown in Figure 11, the general consensus across

---

[25] Other response option included: "NIST is a U.S. leader in cement and concrete academic research" and "NIST is an average contributor to U.S. cement and concrete academic research."

these responses shows that the majority of respondents, in each case at least 70 %, find HYPERCON work on sand and particle shape to be either "highly applicable" or "somewhat applicable" to their own work.

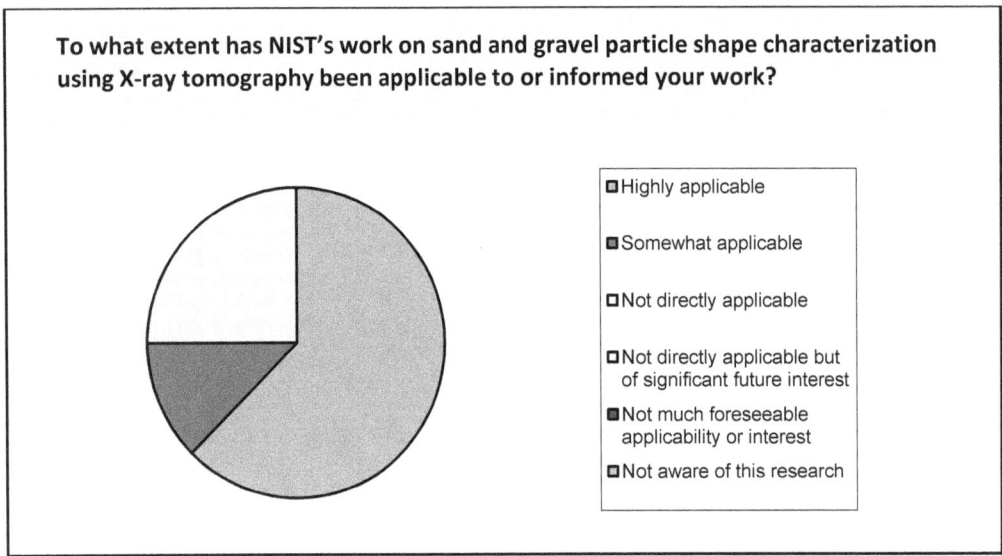

*Figure 11. Aggregates Survey Response: Applicability to Industry Stakeholders*

Responses for other areas of interest to aggregate producers, such as rheology and increased service life were mostly related to the research activities within the industry itself; however, the responses for internal curing were focused around marketing activities. It is notable that the majority of respondents demonstrated "interest in the concept of internal curing, but are not aware of NIST efforts in this area." 28.6 % "are currently marketing materials for internal curing and have utilized NIST tools such as the equation or Monograph for mixture proportioning." This disparity in responses based on awareness of HYPERCON-enabled concepts demonstrates the need for industry-ready research to be "marketed" more rigorously to industry. This was a generalized finding for other technical areas focus and across industry stakeholders.

Survey respondents were asked their opinion concerning the extent to which HYPERCON research "has met the needs of the aggregates industry over the last eight years." As shown in Figure 12, one-third responded that "relevance has increased in recent years, while one-half state that HYPERCON research has been "highly relevant" and maintained at that level throughout the study period.

It is worth noting that in open-ended explanations, the scientific value of HYPERCON research was lauded, but it was routinely explained that in a general sense (the aggregates) industry does not recognize its potential. The following response from an aggregates industry representative summarizes this finding: "NIST is a great institution. The inorganic materials group is excellent, but industry does not seem to appreciate how their work may improve the materials we all use every day (and the fact that they may make profits from the improvements)."

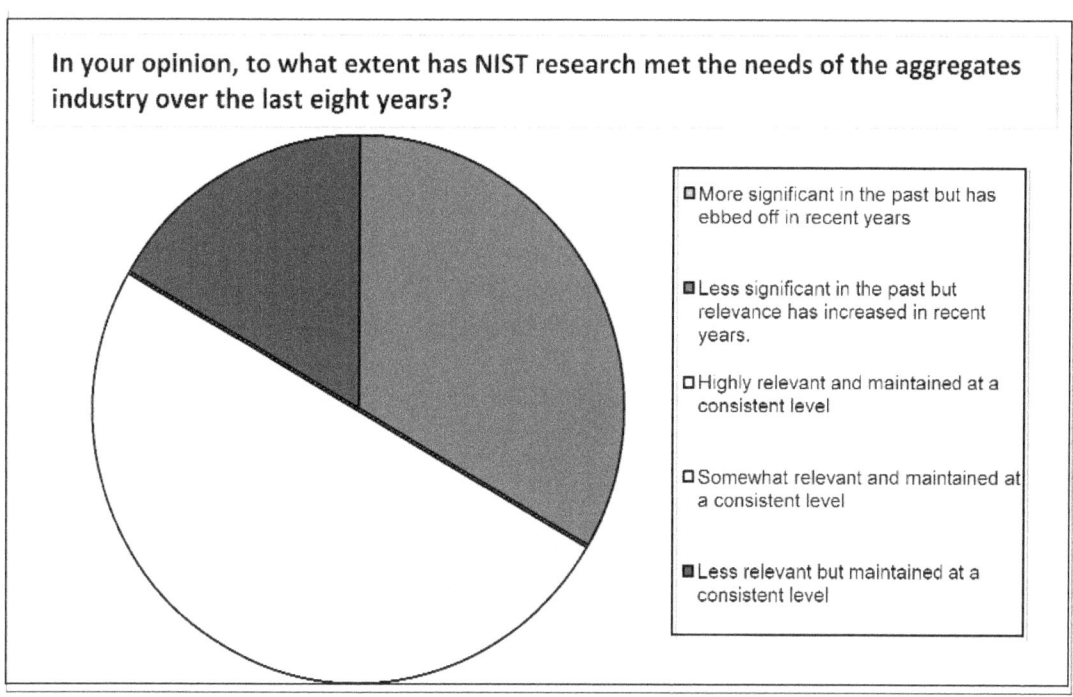

In your opinion, to what extent has NIST research met the needs of the aggregates industry over the last eight years?

- More significant in the past but has ebbed off in recent years
- Less significant in the past but relevance has increased in recent years.
- Highly relevant and maintained at a consistent level
- Somewhat relevant and maintained at a consistent level
- Less relevant but maintained at a consistent level

*Figure 12. Aggregates Survey Response: Significance of HYPERCON Impact*

The following potential research areas were identified by respondents as important to the aggregates industry in the next five years should they be adopted into or strengthened within the HYPERCON research program:
- o Development of models to optimize particle packing of real aggregates
- o Characterization of "super fines" and their impact on rheology and finish ability
- o Minimization of fines generated when crushing coarse aggregates
- o Rapid and reliable measurement of density of granular masses
- o Continuous, real time measurement of aggregate properties, including moisture, size, shape, and texture
- o Elimination of shrinkage and creep in concrete

5.2.4 Chemical Admixtures Survey Responses

Generally, chemical admixture companies surveyed are also part of the VCCTL consortium. To this point, most of the responses were directly related to and correlated with questions concerning VCCTL research and management, as discussed in Section 5.2.12.

The survey responses within the chemical admixture industry followed diverse patterns of satisfaction with HYPERCON research. This is captured by the fact that 25 % of respondents found that HYPERCON "has not significantly met the needs of the chemical admixture industry," while 75 % found that the "relevance of [HYPERCON research] has increased in recent years." This was contrasted, as is suggested by grounded theory, with consideration of the applicability of HYPERCON research in development, production, and marketing of chemical

admixtures. The responses were split 50-50 between HYPERCON research being "somewhat applicable" for these uses and "not directly applicable, but of significant future interest."
To this point it is important to find where the disconnect comes between project research and applicability to the chemical admixture industry. Specifically, 50 % of respondents have not found HYPERCON work on cement hydration and 3-D microstructure modeling directly applicable to their work. Yet, 25 % find is "somewhat applicable," and an additional 25 % find it "not directly applicable, but of significant future interest." These are the areas of research which are the most new to the HYPERCON project portfolio. Thus, they appear to be following the same pattern of stakeholder interest based on their future potential applicability.

The following potential research areas were identified by respondents as highly benefiting the chemical admixtures industry in the next five years should they be adopted into or strengthened within the HYPERCON research program:
- o Cement/admixture compatibility
- o Prediction of strength development of concrete with use of chemicals
- o Prediction of concrete properties with use of blended cements
- o Prediction of strength development of concrete with use of chemicals
- o Prediction of rheology with use of chemicals

5.2.5 State Departments of Transportation Survey Responses

In recent years state DOTs and their associated testing laboratories have taken an increased interest in HYPERCON research in general and a role in VCCTL more specifically.
Those surveyed were asked "to what extent has NIST research met the technical needs of your State DOT over the last eight years?" As shown in Figure 13, fifty percent of respondents found that it has been "highly relevant and maintained at a consistent level throughout the period." Just over thirty-three percent found that it was "somewhat relevant," and 16.7 % found it to be "less relevant and maintained at a consistent level." This question was contrasted by one investigating the extent that HYPERCON research has met the technical needs and challenges presented by cement/concrete in general in the past eight years. These responses were almost identical. As shown in Figure 14, the majority indicated that it has been "highly relevant and maintained at a consistent level throughout the period."

Of specific interest to State DOTs is the development of new forms of concrete (e.g. self-consolidating and pervious concrete). When asked about the role of HYPERCON research in this pursuit, 50 % found it "somewhat relevant and maintained at a consistent level." 25 % found it "less significant in the past, but relevance has increased in recent years;" the remaining 25 % found it "highly relevant and maintained at a consistent level." Respondents were divided strongly among those very familiar with NIST's work in the area and those that do not recollect such work. An anecdote of successful HYPERCON research implementation comes from the state of Florida, where the "use of internal curing of concrete has allowed the FDOT to use more lightweight bridge decks and increase the capacity of the structure without requiring a complete bridge replacement. This has saved millions of dollars for the state of Florida." Others responded that they are "not familiar with what work NIST has done in this area" and that they are "not aware of significant research by NIST in the area."

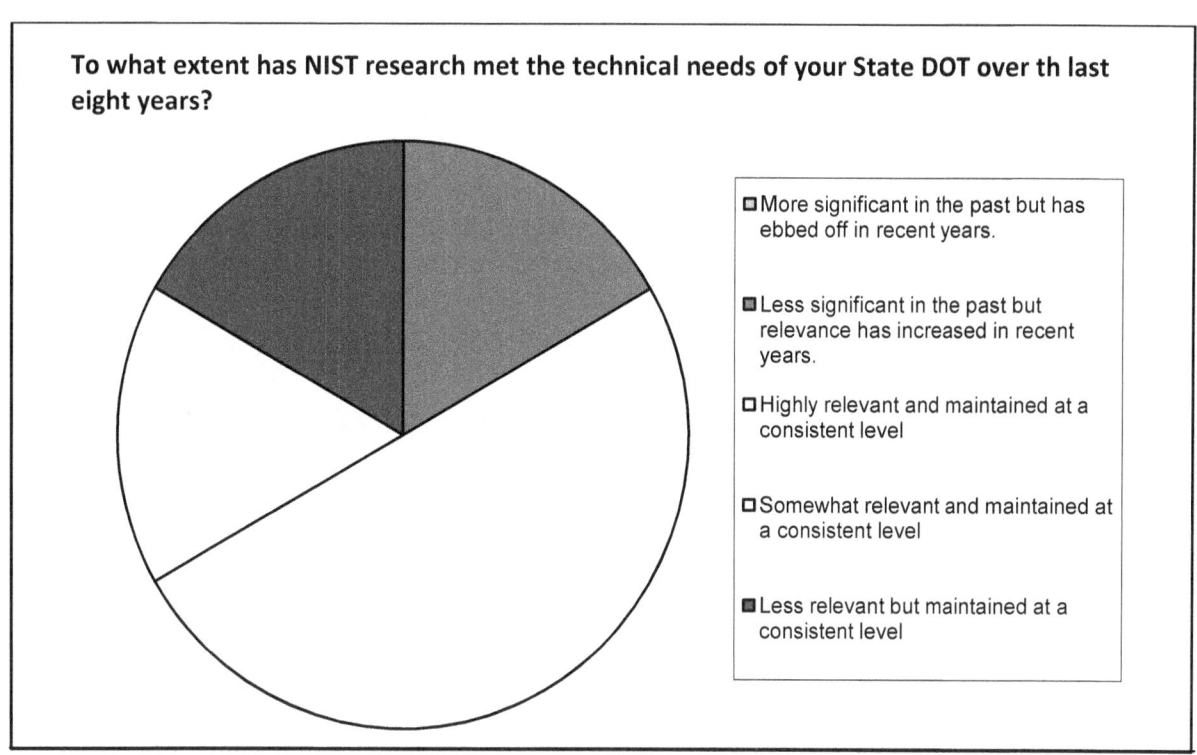

*Figure 13. State DoT Survey Response: Relevance of HYPERCON to DOTs*

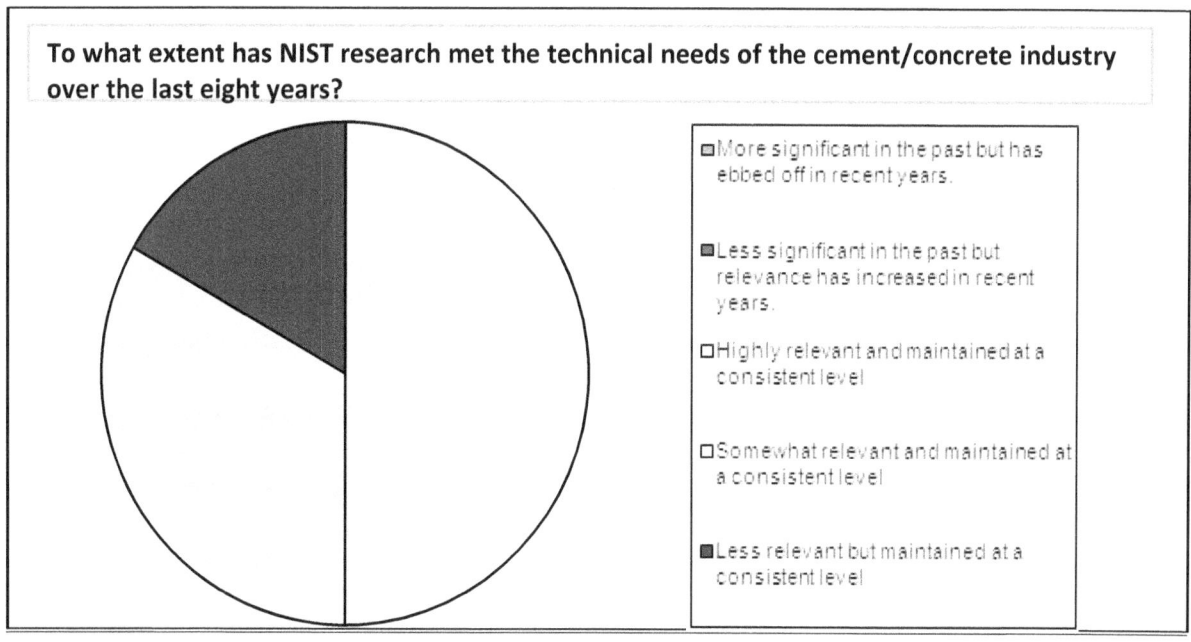

*Figure 14. State DoT Survey Response: Relevance of HYPERCON to Cement/Concrete Industry*

Additionally, State DOTs differ greatly in their levels of expertise and the extent to which they deal with technical issues, such as moisture proportioning. Many tend to ignore software options and use in-house spreadsheets, while others have their mixture proportioning done entirely

55

through contractors. Yet, there was an acknowledgement of the use of in-house software with the aid of HYPERCON-developed software. This diversity within a stakeholder group highlights the challenge to the HYPERCON program to create a method of transmission for research that can be applied by a stakeholder group that varies so widely in its use of such research.

Yet, it is evident that HYPERCON has taken significant steps to satisfy the specific needs of State DOTs. For example, one respondent indicated that "although we have been unable to attend the Workshop and other training, we were able to send a chemist to NIST for a week. He was able to learn from [a HYPERCON researcher] many direct measurement techniques. It was a very valuable opportunity for my agency." Finally, respondents demonstrated a desire to work closely with NIST to determine the best approaches to solving research needs in their work. "It is good to see a survey directed to the DoTs and all the great work done at NIST to assist [State DoTs] (especially during this hard economic time.)"

The following potential research areas were identified by respondents as highly benefiting State DOTs in the next five years should they be adopted into or strengthened within the HYPERCON research program:
  o Direct measurement of Bogue compounds and change in specifications resulting from direct measurement.
  o Ability to predict the long term durability of the raw materials being used in the structure and then being able to validate that model at an early age (less than ten years).
  o Increased research on service life and materials characterization.


5.2.6 Ready Mixed Survey Responses

In order to gauge the level of technical knowledge among respondents to the ready-mixed survey, they were asked to indicate their use of commercial or in-house software for mixture proportioning and optimization. Fifty percent of respondents use in-house software as part of a "quality management system," while the remainder use other less technical means for proportioning. There is a strong desire for those not currently using software to move towards its use, especially if it enables better strength modeling and the capability of "smart" durability calculators.

The survey sought to determine the level at which HYPERCON research has met the needs of the ready mixed industry over the past eight years. To this point, 60 % of respondents found that HYPERCON research has been "less significant [to their work] in the past, but relevance has increased in recent years." Twenty percent disagreed; they found HYPERCON research "more significant in the past, but the relevance has decreased in recent years." The 20 % balance of respondents found HYPERCON research to be "somewhat relevant and maintained at a consistent level." This disparity in responses demonstrated a variation in how different actors in the ready mixed industry interact with and use HYPERCON-generated research. This variation is also evident in responses for the specific technical research areas which are strongly related to the ready mixed industry.

The following potential research areas were identified by respondents as highly benefiting the ready mixed industry in the next five years should they be adopted into or strengthened within the HYPERCON research program:

- o Air entrainment stability in the presence of carbon
- o Better predictability of cement/admixture incompatibilities
- o A "signature" state of concrete for acceptance in a "fresh state"
- o Handling/management of stack emissions
- o Concrete cracking-related distress mechanisms

The following discussion is broken down into research areas of common concern among stakeholder groups. These research areas are analyzed in this fashion in order to highlight how HYPERCON research has dealt with the diverse needs of different stakeholders over the same research area. The common research areas include: internal curing, rheology, x-ray diffraction, and prescription to performance.

5.2.7 Internal Curing Research Program Responses

Internal curing of concrete is an issue of particular concern to aggregate producers, State DoTs, and the ready mixed industry. The results among these three surveyed groups for the level to which HYPERCON research on internal curing has been utilized is reported in Figure 15. Among the ready mixed respondents, none have utilized HYPERCON tools for work on internal curing, while 1/3 of State DOTs had done so and 29 % of aggregates producers. An interesting finding is that across the three surveys, 37 % of respondents do not have knowledge of HYPERCON research efforts on the topic, though they are interested in learning more about internal curing for their own research pursuits. This finding highlights an opportunity for improved similar lacks in communication to industry about the research available to them through HYPERCON.

Those that have used HYPERCON-enabled internal curing techniques are very pleased and report significant cost savings from the effort. For example, one aggregates producer reports that: "We have installed over 500,000 cubic yards of internally cured concrete. The efforts from NIST have made this possible; the tireless efforts toward investigating and discovering the science behind the mitigation of cracking for concrete is terrific. This profound discovery has and will be invaluable to infrastructure developments in the U.S. and the world. We have the opportunity to save billions of dollars by increasing the longevity of structures via internally cured concretes; therefore, funding needs to increase rather than decrease. Additionally, by increasing longevity we are also reducing our carbon footprint."

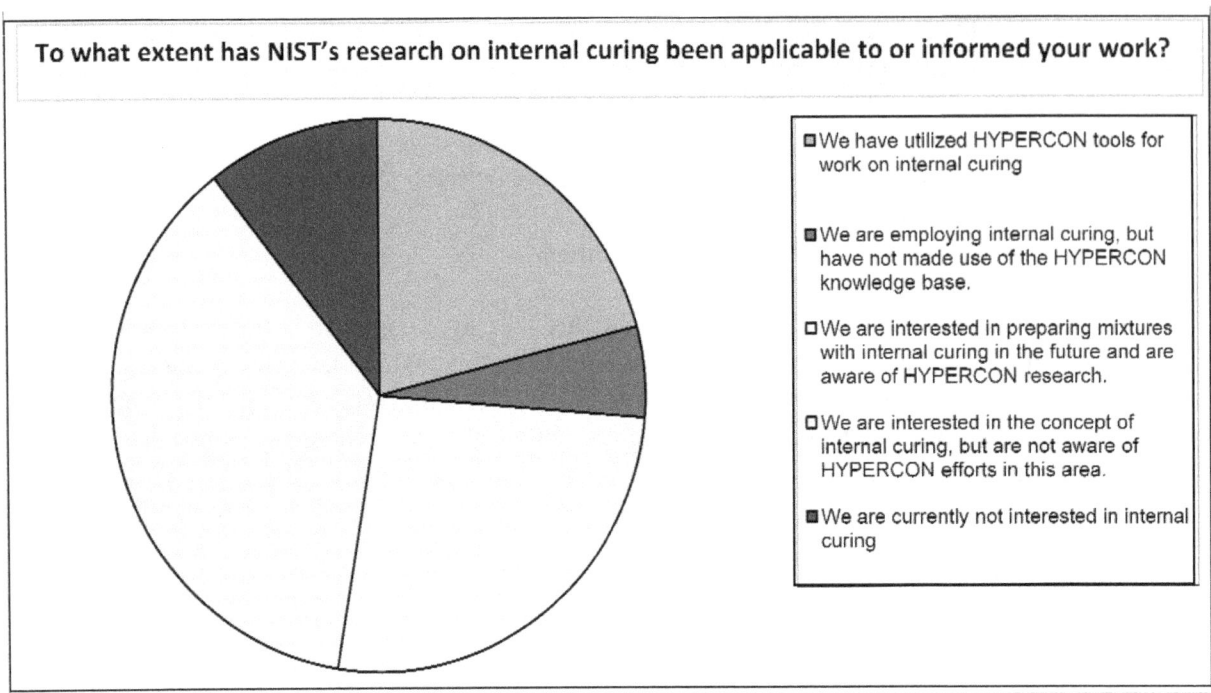

*Figure 15. Aggregates, State DoT, and Ready Mixed Survey Responses: Usage of HYPERCON Internal Curing Research*

5.2.8 Rheology Research Program Responses

Rheological studies are relevant to stakeholders within the aggregates, chemical admixtures, and standards organizations. Within HYPERCON there are two branches of rheological research; one is applied and concerned with real-world measurements, while the second is focused on modeling these real-world attributes without physical testing. The three relevant stakeholder groups were asked to what extent "HYPERCON work on rheology/processing using developed metrological methods for accurately measuring and modeling the rheology of cement paste, mortar and concrete has been applicable to or informed" their work. As shown in Figure 16, no chemical admixture company or standard setting body responded that the work had been "highly applicable;" however, 38 % of aggregates producers responded that it had been. Twenty-two percent of respondents indicated that to date the research "was not directly applicable, but that [they have] significant future interest in it." Finally, 38 % find "little applicability of future interest in this brand of research." Upon investigation, it seems that respondents familiar with this research feel that it is still in its infancy, and should such research truly lend itself to standardization between rheometers, for example, they would have a much greater interest in it.

The second branch of rheology research is concerned with modeling of concrete rheology/processing using real sand and gravel particle shapes and inter-particle forces. Respondents to the aggregates and chemical admixture surveys were asked the extent to which such research has been applicable to their work over the past 8 years. As shown in Figure 17, 42 % of respondents found this work "somewhat applicable" to their own work, while 1/3 of aggregated responses indicate that it has been "highly applicable." Finally, the ¼ of respondents

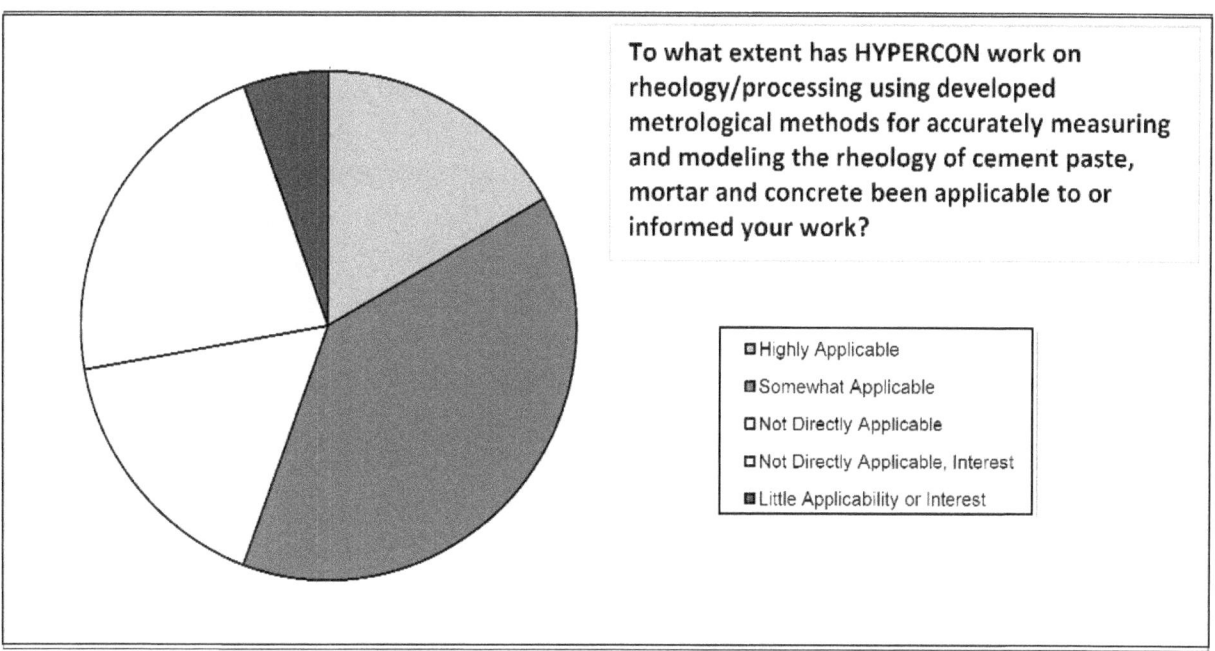

To what extent has HYPERCON work on rheology/processing using developed metrological methods for accurately measuring and modeling the rheology of cement paste, mortar and concrete been applicable to or informed your work?

□ Highly Applicable

■ Somewhat Applicable

□ Not Directly Applicable

□ Not Directly Applicable, Interest

■ Little Applicability or Interest

*Figure 16. Aggregates, Admixtures, and SDO Survey Response: Usage of HYPERCON Applied Rheology Research*

that found the research "not directly applicable" were aware of it and have "significant future interest" in such research. This pattern of all respondents having good knowledge of NIST research in this ICME realm is a break with other patterns of stakeholder groups being unacquainted with specific HYPERCON research that is generally highly important in their field. To some extent this might have to do with the extent to which ICME methods are sought in the field as well as the significant overlap between these stakeholders and members of the VCCTL consortium.

5.2.9  X-ray Diffraction Research Program Responses

X-ray diffraction methods are employed by stakeholders in the chemical admixture, ready mixed, and State DoT stakeholder groups. These respondents were asked the extent to which "HYPERCON's work on cement characterization using X-ray diffraction, scanning electron microscopy, and optical microscopy has been applicable to or informed your work?" As shown in Figure 18, in the aggregated sample, 62 % of respondents expressed that this work had been highly or somewhat applicable throughout the past eight years. The rate for State DoTs was at 80 %. The 8 % of respondents finding "not much foreseeable interest or applicability" are from industry organizations, representing general interests of individual groups and companies, but not directly their members' in-house research activities.

59

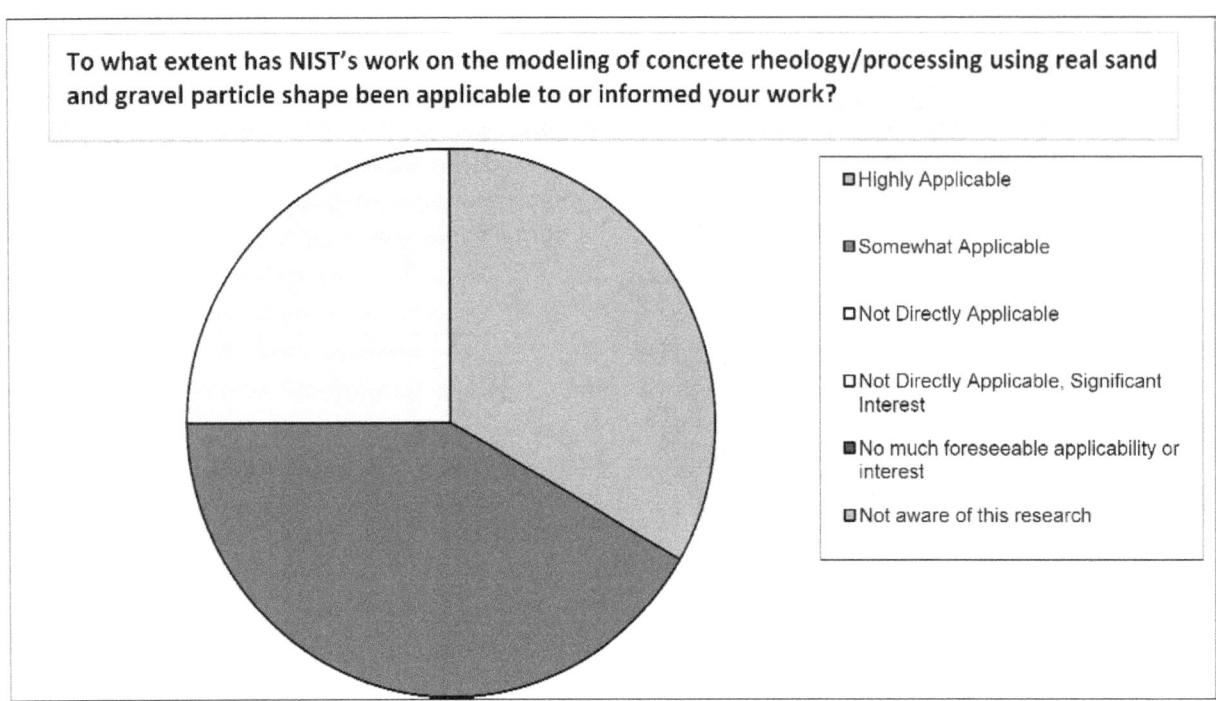

*Figure 17. Aggregates and Admixtures Survey Response: Usage of HYPERCON Modeling Rheology Research*

5.2.10 Prescription to Performance Research Program Responses

Prescription to Performance (P2P) has been described by the National Ready-Mixed Concrete Association (NRMCA) as a desire to change concrete specifications from a prescriptive form to a performance form, which in turn, drives R&D for proprietary advantage in the marketplace. This is a big challenge for industry because workable performance specifications require a greater level of materials science understanding and performance prediction in standards tests. In recent years the P2P initiative has carried significant weight in the ready-mixed industry, which accounts for 75 % of all concrete produced. This issue is also of some interest to the State DoT stakeholder group due to the high amounts of concrete used and the significant maintenance for which they are responsible. As shown in Figure 19, 37 % of responses from the ready mixed and State DoT stakeholder groups combined find that HYPERCON research in this area has been "less significant in the past, but relevance has increased in recent years." Thirty-six percent find that HYPERCON research concerning P2P to be "somewhat relevant and maintained at a consistent level." These findings suggest the need for more communication with these stakeholder groups to better define HYPERCON's role in fulfilling their P2P research needs.

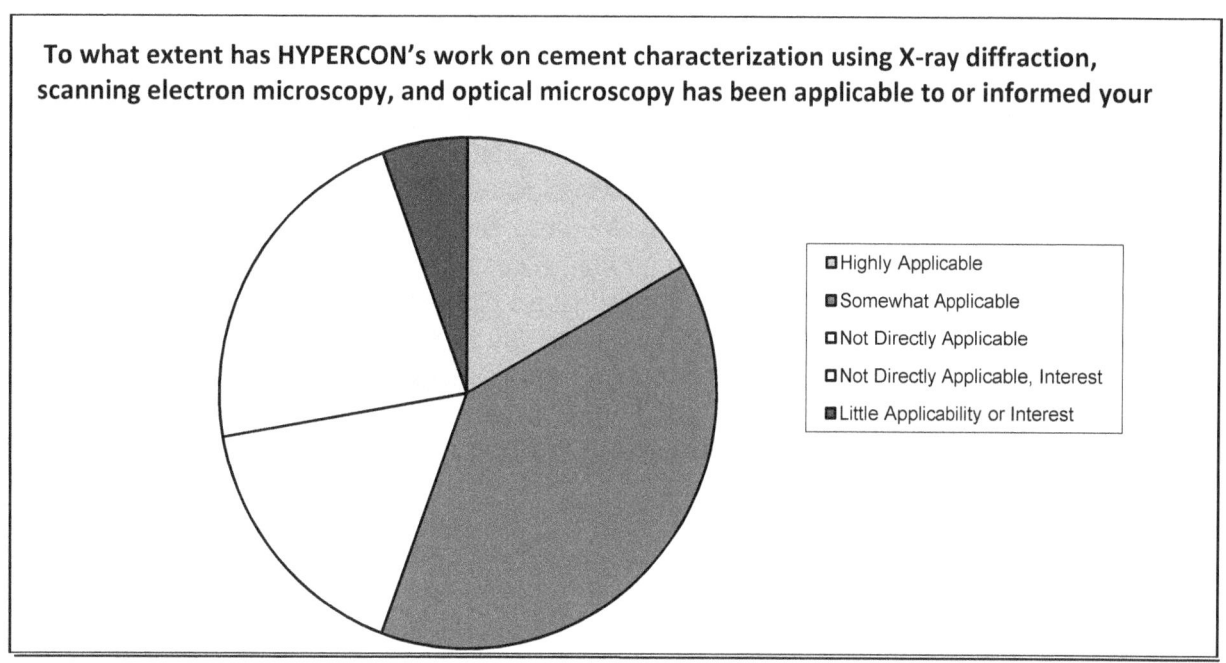

*Figure 18. Admixture, Ready Mixed, and DoT Survey Responses: Usage of HYPERCON X-Ray Diffraction Work*

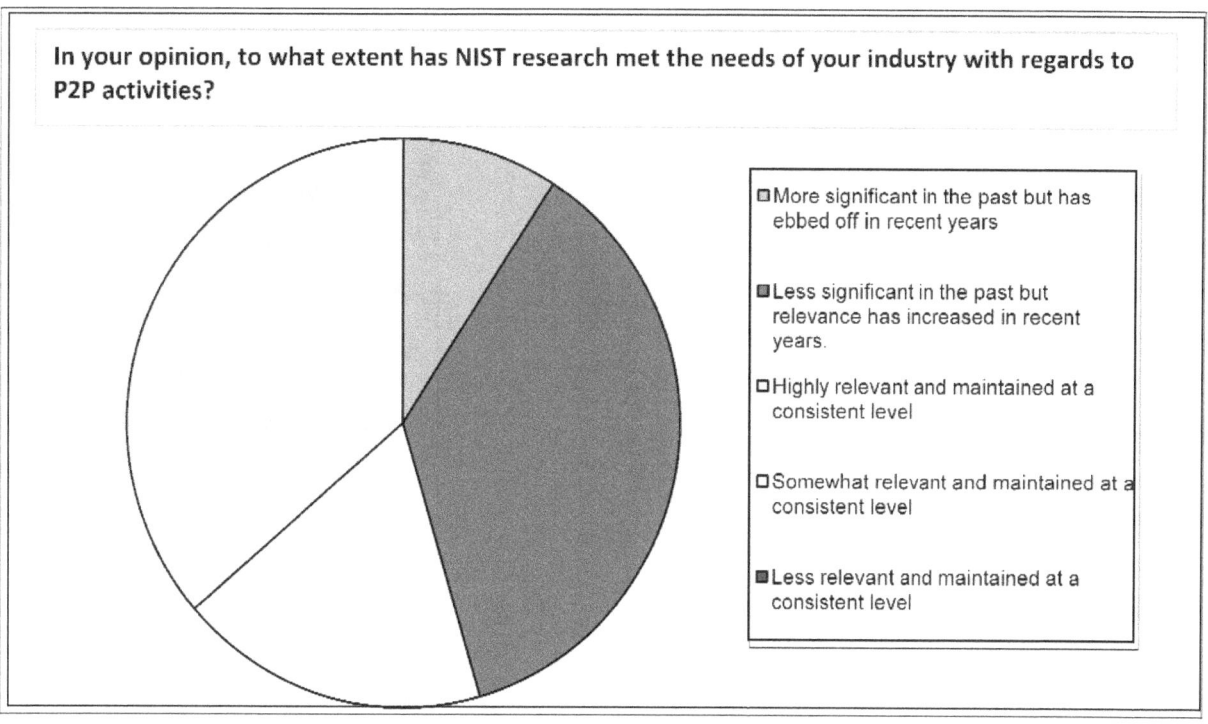

*Figure 19. Ready Mixed and DoT Survey Responses: Relevance of HYPERCON P2P Research*

## 5.2.11 VCCTL Research Program Responses

VCCTL is well-known throughout the cement/concrete industry. This major product of the HYPERCON program is designed to address complex research issues, such as transforming towards a performance prediction measurement science platform. Within each of the stakeholder surveys there were questions concerning awareness and use of VCCTL. Answers are considered from the perspectives of aggregates, chemical admixture, State DoTs, and ready mixed stakeholders. Of these respondents, 8 have at one time been a member of the VCCTL consortium, while 24 have not. Among those that have never held a membership, 90 % have "not nor have ever been members of the VCCTL consortium, but are very interested in the further development of the VCCTL software." As shown in Figure 20, no members of VCCTL report significant use of the software product, but 75 % do report "limited use of the software," opposed to the 25 % of members that report no use of the software.

These responses indicate a dual challenge for HYPERCON researchers to determine: (1) the aspects that will move usage of members from "limited" to "significant" and (2) software features that will entice those interested in the software but not yet part of the consortium to join. These challenges are magnified in an economically difficult time for the industry. Many responses indicated that "HYPERCON is doing leading edge research; VCCTL has great potential for solving industry problems. Unfortunately, research funds have been exhausted and without more tangible results, we cannot stay in the consortium."

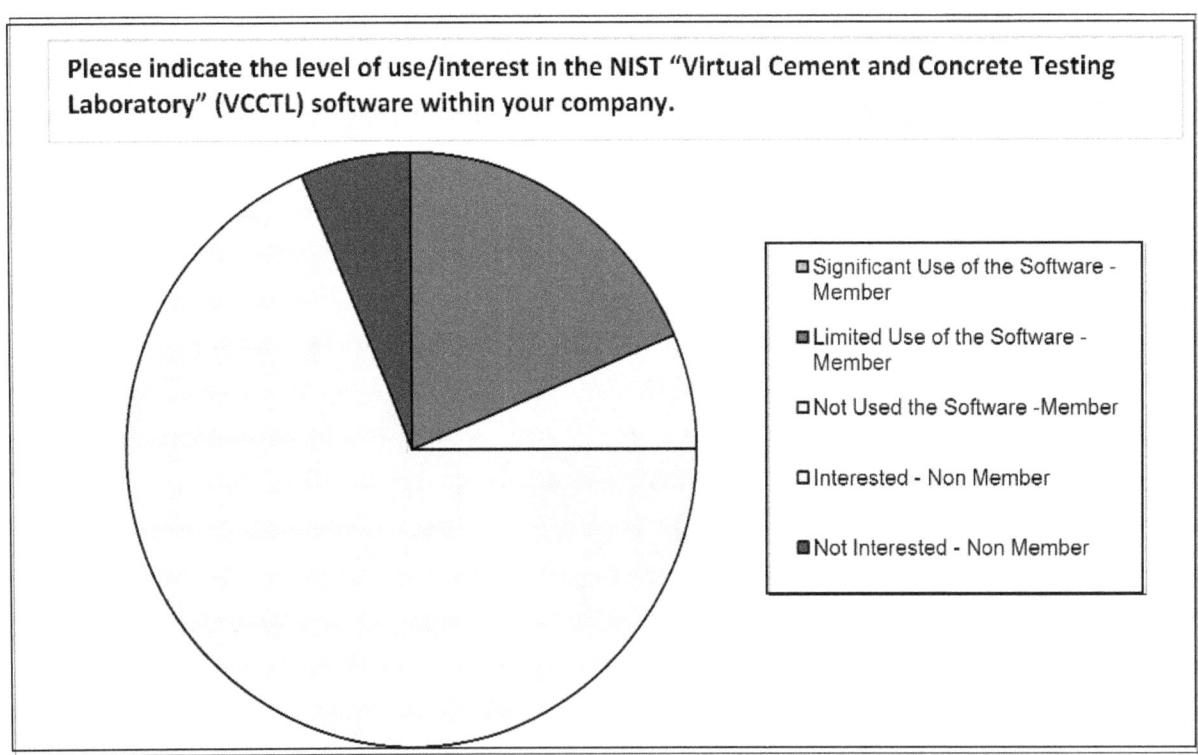

*Figure 20. Aggregates, Admixtures, DoT, and Ready Mixed Survey Responses: Usage of VCCTL Software*

The next subsection provides a more detailed discussion of VCCTL from the perspectives of present and past consortium members.

5.2.12 VCCTL Survey Responses

This subsection reviews responses to the surveys of current and past VCCTL consortium members. This review provides an understanding of the benefits of membership as well as a basis for comparison with other consortia activities within the industry, as discussed in the subsection that follows.

***Current VCCTL Members***
The majority of current VCCTL members, as of February 2009, are classified as chemical admixture companies or have a strong interest in admixtures. Of these, one works for a U.S.-owned entity, while the rest are under foreign ownership, but with U.S.-based activities. There was as little overlap as possible between respondents to this survey and respondents to the chemical admixture stakeholder survey; one respondent contributed to both. 2 companies joined VCCTL upon its inception in 2001, while 1 joined in 2003, another in 2004, and finally two more in 2008.

Respondents were asked to identify the main research areas they would like to see the VCCTL consortium explore in the future. All members agreed on rheology and hydration as two compelling research topics. Fifty percent of respondents are interested in heightened research concerning mechanical properties. The challenge of designing the research agenda for VCCTL, even when members are from the same stakeholder group, is demonstrated by the diverse interests shown. One-sixth of respondents was interested in more research in each of the following general topics: mix design, placement, curing, and service life.

An underlying purpose of the VCCTL research is to reduce physical testing, which can take upwards of 28 days for a single sample. As shown in Figure 21, however, it is telling that no member responded that the VCCTL software had had either "significant impact" or even "somewhat significant impact" on their physical testing work. Though, 49 % do find that there is a "significant potential application" in the future. Rather than a limitation in the VCCTL software itself, this likely highlights the naissance of the ICME approach, for which VCCTL is a leader. Respondents were asked the extent to which the VCCTL software package, aside from general VCCTL research findings has informed and been applicable to the stakeholders work. Once again a large share (50 %) of respondents indicates that it has "not been directly applicable, but is of "significant future interest" to their company. One-third responded that it has been "somewhat applicable" to their work. Sixty percent find the VCCTL software interface "acceptable" to navigate, while 20 % indicate that the "interface could use minor improvements" to make it more user friendly.

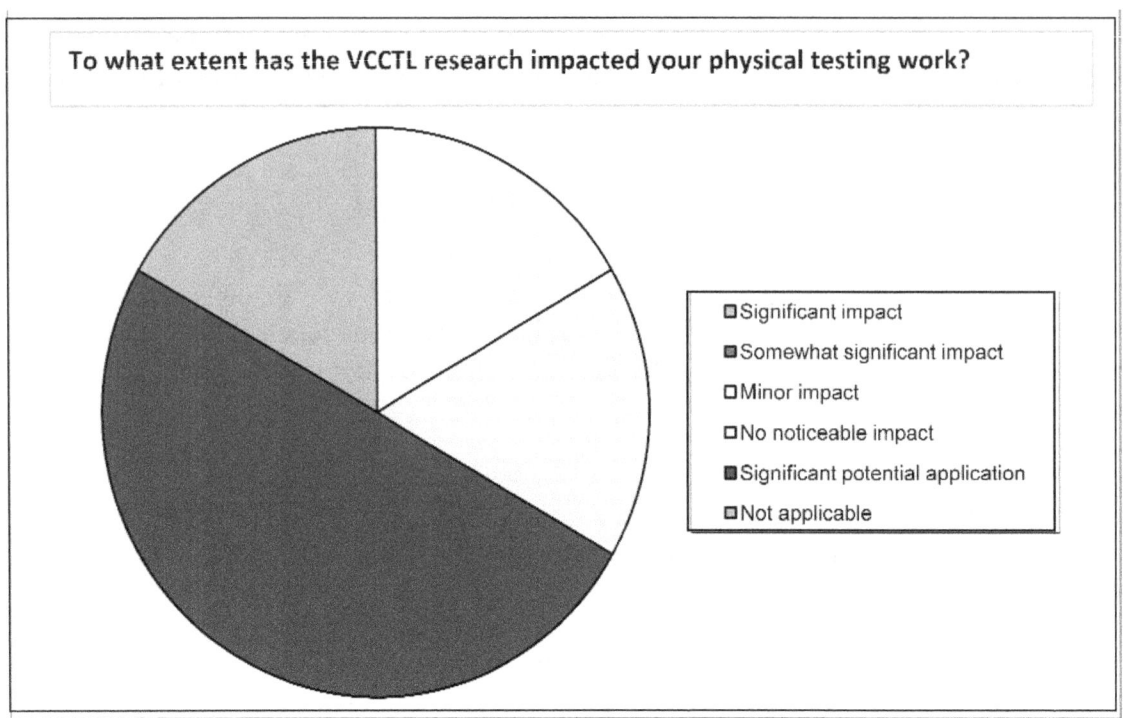

*Figure 21. VCCTL Members Survey Response: Significance of VCCTL Research Impact*

The survey explored the applicability of specific areas of VCCTL research to determine their overall significance to consortium members. In these cases the response patterns followed closely from those observed for the overall HYPERCON program. When asked "to what extent work on modeling of concrete rheology/processing using real sand and gravel particle shape and inter-particle forces has been applicable to or informed work," 66.7 % responded that it was "not directly applicable, but of significant future interest." While 16.7 % responded that it was "highly applicable" and the remaining 16.7 % described it as "not directly applicable." In comparison with other research topics, there was a large jump in significance reported for VCCTL cement hydration and 3-D microstructure modeling. Fifty percent of respondents rated this work as "highly applicable," while 33.3 % rate it as "not directly applicable, but of significant future interest."

Throughout the survey there were few responses indicating that the VCCTL research was indeed of significant interest or applicability to the daily research and marketing work taken on by the member companies. It is recognized that companies became members of the consortium at different times; newer members may have limited knowledge of the research and VCCTL software. While, the most frequent response patterns demonstrate a very limited impact across all members, there is still a belief that future VCCTL research will have significant impact on the industry. Respondents were asked to indicate their company's main motivation for maintaining consortium membership. Their responses follow:
   o Technology exchange
   o Gain basic insights into rheology and hydration. This provides additional thinking tools when facing an internal consulting question as a central R&D facility for our company
   o Interest in modeling cement hydration and rheology

o Fundamental research
o Technology leadership
o Participating in one of the advanced groups in modeling of cement hydration

Furthermore, 50 % of respondents indicated they "are satisfied with the VCCTL research output, but that it could be more geared towards [their] specific research needs." 16.7 % of respondents state that "everything in the VCCTL research program is of use to [them]," while the remaining 33.3 % state that "only some of the VCCTL research is meaningful to us." When asked about continuing membership into the coming year, 1/3 of respondents were confident in their desire and ability to do so, while 2/3 were "not sure" based on the "economic crisis and associated budget restrictions."

There was a general consensus that the semi-annual VCCTL meetings are useful. The meetings provide direct exchange of ideas and communication. Those traveling from as far as Switzerland to NIST for these meetings recognize that they allow a platform for a good understanding of the expectations each company has for consortium output and vice versa.

The interplay between companies' in-house research activities and those that are undertaken directly as part of VCCTL is complex and varies significantly from consortium member to consortium member. One-sixth of current members had pre-existing collaboration activities with HYPERCON before joining the VCCTL consortium. Forty percent of members now work directly with a NIST researcher in order to augment the VCCTL research activities; none work collaboratively on HYPERCON research that does not directly feed into VCCTL. Finally, 60 % participate in VCCTL as "passive" members, providing no direct collaboration to reach the agreed upon research goals. Effectively, these members are active participants in setting the research agenda for VCCTL, but do not intend to participate with NIST in its fulfillment. Consortium members were asked what direct benefits they have seen from VCCTL membership. The responses follow:
o Help in educating interning students
o Exchange of highly technical knowledge
o Basic knowledge of rheology and hydration. VCCTL provides the only software world wide that gives true basic insight needed as a part of our central R&D resources.
o The modeling, though imperfect, can finalize results taken from traditional analytical tests

There are a number of changes and additions to the VCCTL research agenda that would help consortium members in their own research:
o Impact of admixtures on rheology
o Impact of admixtures on hydration
o Reactivity of blended cements
o Modeling in cement hydration
o Less "CO2" cementitious material utilization
o More effective use of manufactured aggregates
o Effect of different sands on rheology (especially in the realm of prediction)
o Effect of different cement types on hydration and rheology
o Study and control of hydration time

The range of VCCTL direct member benefits and research interests, while informative in its own right, also indicates that the consortium has been successful in creating an active and healthy exchange of ideas.


### Past VCCTL Members

The opinions of past VCCTL consortium members concerning the VCCTL research foci is now reviewed. Of the 6 respondents to this survey, 2/3 are from the cement production stakeholder group, 1 represents an aggregates company, and the remaining 1 represents a chemical admixture company. This mix of stakeholder groups demonstrates that VCCTL membership has become more confined to one stakeholder group over time. This seems to be the product of changes in VCCTL direction over time, especially with respect to software evolution, as well as recent financial concerns within the cement/concrete industry. One third of these respondents work for a U.S.-owned company while the rest are foreign-owned entities. Two thirds of the respondents became members of VCCTL in its inaugural year, 2001; 1 in 2002 and 1 in 2003. Starting in 2004, one company left the consortium each year from 2004 through 2007. Two companies left in 2008.

Research topics of interest during the period of membership in VCCTL are highly varied and well supported among this group of respondents, as shown in Table 13.

*Table 13. Past VCCTL Member Survey Response: Research Topics of Interest*

| NIST Cement/Concrete Research Program (VCCTL--past) | | |
|---|---|---|
| **Which of the following topics were you interested in during the time of your membership in VCCTL? (check all that apply)** | | |
| Answer Options | **Response Frequency** | **Response Count** |
| Mix Design | 50.0% | 3 |
| Placement | 16.7% | 1 |
| Curing | 33.3% | 2 |
| Mechanical properties | 66.7% | 4 |
| Hydration | 83.3% | 5 |
| Rheology | 50.0% | 3 |
| Service Life | 33.3% | 2 |
| *answered question* | | 6 |
| *skipped question* | | 0 |

The response pattern concerning the applicability of the VCCTL software package is comparable for this group to that of present VCCTL members. No respondents found that the software package was "highly applicable" to their work, but 50 % did find it "not directly applicable but of significant future interest" and 17 % answered that it was "somewhat applicable." Seventeen percent found that VCCTL work "significantly impacted" their company's physical testing work; 50 % report "no noticeable impact," and 33.3 % recognize a "significant potential impact." Specifically, when asked the extent to which "work on the modeling of concrete rheology/processing using real sand and gravel particle shape and inter-particle forces was

66

applicable to their work," 1/3 responded "not directly applicable" and the remaining 2/3 stated that it is "not directly applicable but of significant future interest." With regards to the applicability of VCCTL cement hydration and 3-D microstructure modeling, 50 % found it to be "not directly applicable, but of significant future interest."

The VCCTL software interface was a stumbling block for some of the respondents. 50 % report that the interface was "acceptable," but 1/3 of respondents state that "the interface was very difficult to use and should have been restructured." Though current users did report the desire for "minor improvements" to the interface, the trend seems to show that over time the interface has improved substantially.

The survey sought to determine the motivation for the companies to join and then terminate membership as part of VCCTL. They were asked the extent to which VCCTL research was geared to their companies' needs. Responses were divided evenly among: "yes, everything in the VCCTL research program was of use to us;" "we were satisfied with the output, but it could have been more geared towards our specific research needs;" and "only some of the VCCTL research was meaningful for us." No responses reflected that "VCCTL research outputs were not at all useful for [the] company's needs."

As noted, the majority of these respondents became VCCTL members in its first year and served to outline the consortium's basic research agenda. Respondents were asked to state their main motivation for initially joining VCCTL; representative responses follow:
- o  Saw VCCTL as a future tool to help industry solve problems and develop new technologies and high performance materials
- o  To be involved in more accurately modeling of aggregate shape and to become familiar with the system
- o  To obtain new tools in order to reduce tests; to develop competence in modeling research and durability testing
- o  Gain insight to the hydration process
- o  Use a hydration model to be able to predict properties of cement-based materials; particular interest in being able to predict properties in the presence of mineral additions

It is evident that much of continued membership in the VCCTL is based upon expectation of future research outputs. As a cutting edge and standalone platform for ICME in the industry, consortium membership will help entities take a competitive position once VCCTL makes major breakthroughs. However, members have not changed their view of VCCTL as a strictly "exploratory" research enterprise since its inception. Eight years later, members tend to maintain membership for the same reasons they had in the past, effectively in expectations of "significant applicability" of VCCTL findings in the future. 50 % of respondents do report continued collaboration with VCCTL researchers outside of the consortium since they left.

All past members reported benefits from their membership in VCCTL. As opposed to current members, 50 % of past members cited "significant help in educating students" as a primary benefit. Other benefits were:
- o  Streamlined testing
- o  Expedited design and development of new materials

- o Drove changes in standard test methods
- o New technology development
- o Spurred in-house research on new materials and durability

Past members of VCCTL were asked why they left the consortium. Those that left the consortium recently cited the economic downtown and resulting lack of funds for such an activity. Yet, even among those leaving the consortium before the current global economic slowdown, two indicated that as internal budgets were restricted, VCCTL membership was one of the first research activities they were forced to cut. Of the respondents, 1/3 would like to reinstate membership in VCCTL if given the opportunity in the future. Fifty percent were unsure if they would, and 17 % stated they are not interested in rejoining the consortium. Those who are not certain of their interest in future membership cited the following weaknesses of VCCTL: "the project was not easy to follow, computer code was only understandable to insiders, little progress over many years. In the end it was too much money for little benefit." Another observation was "invested an important amount of money, but really was not obtaining the usefulness in terms of being able to apply the research to real life situations." Yet, these respondents recognize that VCCTL provided information that may have been more useful had they followed the project more closely. Those who left the consortium some time ago would like to have the opportunity to reassess the model and determine its current value to their company.

Effectively, it appears that the cement/concrete industry as a whole has a great deal to gain from ongoing VCCTL research; however, individual companies have questioned their viability as a continued source of its funding. There seems to be a point at which, if in the company's view it was not realizing reasonable return on their membership dues, it felt compelled to drop out of the consortium. It is important that VCCTL serve as a meeting place for different types of stakeholders within the cement/concrete industry; however, the present group of mainly chemical admixture companies may provide VCCTL the clarity of vision and mission that ultimately results in very well-defined and well-received, research outcomes. Additionally, VCCTL may want to consider ways to better encourage members to contribute elements of their company's own research to strengthen and expedite VCCTL outcomes. This paraphrased response summarizes the importance of VCCTL taking on this brand of initiative: "We have enjoyed being part of the VCCTL project and feel sorry that we are not participating any more. We might have put more input into our own work; however, following the project and the kind of slow progress being made, it was difficult for us to stay on board, especially taking the membership fee into account, which was at the high end for the benefit that we received."

5.2.13 Cement/Concrete Consortia Comparisons

In this section each of the cement/concrete consortia introduced in Section 4.2.2 is discussed in detail and compared to defining aspects of VCCTL. Responses reflected here were obtained through a series of detailed interviews with leaders in each consortia as well as informal discussion with members of the various consortia. The goal is to determine the strengths and weaknesses of VCCTL relative to other concrete/cement research consortia. Additionally, potential opportunities for VCCTL to draw on the strengths of other consortia are considered.

*SUMMA*

SUMMA is a private partnership led by Materials Service Life (MSL) LLC and a Canadian Corporation, Simco Technologies,[26] in which Laval University, a Canadian University, is an equity partner. Summa was created in 2003 through MSL in order to address service-life prediction needs addressed in calls for proposals by the U.S. Navy.

MSL has a pre-existing suite of STADIUM software products, which allow users to predict the service lives of saturated and unsaturated concrete structures exposed to aggressive environments (e.g. seawater, deicing salts, and contaminated soils). STADIUM products also allow prediction of moisture transfer and heat transfer within concrete elements and prediction of oil (or diesel) transfer within porous media. The STADIUM software has been used as a backbone for SUMMA software development. SUMMA has added new features, enabling development and monitoring of software programs for construction and maintenance and budgets for evaluated structures. SUMMA's developed software targets owners, engineers, and concrete producers in determining optimum management of service life estimates.

There have been a number of public and private partners in SUMMA's research activities, as follows:
> Public Members:
>> o U.S. Bureau of Reclamation
>> o U.S. Army Corps of Engineers
>> o U.S. Navy
> Private Members:
>> o BASF
>> o Eulclid
>> o Holcim
>> o Lafarge
>> o MMFX
>> o Sika
>> o W.R. Grace

Total investment of the partners in the SUMMA consortium during the period 2003-2009 was $ 5 million, compared with VCCTL member investment of ~ $ 2.3 million over the same period. The activities of SUMMA have recently been officially completed. The produced software is in the process of commercialization. All SUMMA partners will have licenses to use the software. There is an academic version of the software developed by SUMMA for use in university research.

*ACBM (Center for Advanced Cement-Based Materials)*

The ACBM Consortium was established in 1989 as a National Science Foundation (NSF) Science and Technology Center, dedicated to advancement in the cement and concrete industries. NSF funding ended in 2000; since then ACBM has been funded by its private industrial partners. Current private industrial partners are:
> o Haliburton

---

[26] SIMCO Technologies Inc. specializes in the development of numerical models enabling the prediction of the behavior of civil engineering works over short and long periods of time.

- o Holcim
- o Lafarge
- o PCA
- o W.R. Grace

ACBM's efforts are focused primarily on research, education, and technology transfer activities. ACBM Consortium members are broken down into the categories of ACBM Research Institutions,[27] Affiliate Institutions,[28] and Industrial Partners.[29] ACBM as a consortium is geared towards scientists and engineers conducting interdisciplinary research and graduate level education in cement-based materials research. Hundreds of students and visiting scholars have participated in ACBM programs. ACBM partners with NIST to sponsor the annual ACBM/NIST Computer Modeling workshop. ACBM has established interactions with companies, associations, and government stakeholders in cement and concrete technologies.

According to its management, ACBM is facing two main challenges today. First, as potential partners are more concerned over intellectual property, it has created an environment in which research sharing and partnering is more difficult and guarded than in the past. Secondly, the cost of maintaining students continues to increase steadily. ACBM has historically been known for student training and involvement, though without more direct financial partnering and grants this is difficult to maintain. There has been a slight reduction in partnering over time due to a direct decrease in funding for laboratories. Typically sponsors have given between $ 30,000 and $ 35,000 annually for membership fees, dependent upon company size and affiliation level. Yet, due to changes in budgets and industry spending the number of sponsors is down to 3 or 4 contributors annually. ACBM management hypothesizes that this reduction in participation is linked to a concern by U.S. companies for obtaining clear and specified deliverables. Though they agree that strategic basic research is important in moving industry forward, without finalized outcomes on a fixed schedule, investors are becoming more apathetic towards financing such initiatives. It is interesting to note that ACBM management does specify that this attitude is more common among U.S.-owned entities than European ones.

A representative from each affiliate member institution serves on the ACBM Industrial Advisory Board. This Board meets twice annually with central management to review technical progress, structure of projects, and overall research of the ACBM Center.

ACBM faculty has been working with business leaders to develop strategies to implement and enhance the role of innovation in the small scale businesses that primarily populate the cement and concrete industry. Much of the formal interaction in this arena is carried out through ACBM Technology Transfer seminars. These seminars provide a forum for academic-industry discussions and knowledge sharing. Effectively, ACBM seems to identify as a nexus of companies, associations, and universities interested in cement/concrete research efforts.

---

[27] ACBM Research Institutions include Northwestern University, Purdue University, University of Illinois at Urbana-Champaign, University of Michigan, and National Institute of Standards and Technology.
[28] Affiliate Institutions include Universite de Sherbrooke, Icelandic Building Research Institute, Delft University of Technology, Politecnico di Milano, Princeton University, Technical University of Denmark, and Universite Laval. ACBM Industrial partners include: Haliburton, Holcim, Lafarge, PCA, and W.R. Grace.

Projects are intended to lead to a better understanding of the technical needs/priorities of the industrial community. Main ACBM projects deal with chemical additives to reduced concrete shrinkage, use of industrial waste in concrete mix-production, and self-consolidating concrete.

ACBM has been relentless about encouraging and spurring interaction within the cement/concrete community. A main method has been through their publication "Cementing the Future," which is a scholarly and informational newsletter distributed to over 4,000 individuals in the field.

ACBM is focused on relatively short- and medium-term project and research goals. There are overarching ACBM projects and some ICME approaches employed, but since industrial funding has become a main source of funding for the consortium, research focus continues to shift.

### *NanoCEM*

The NanoCEM Consortium is a European Union-based center for research on cementitious materials. NanoCEM was launched in 2004 and strives to "support the development of new and improved materials through fundamental research and knowledge generation." The NanoCEM consortium manages an integrated research and education organization that is focused on fundamental research activities, with a goal to enable technological breakthroughs in the field of cement and its applications. NanoCEM's management credits a great deal of its organizational success to lessons learned through observation of the ACBM consortium.

NanoCEM comprises 24 academic partners and 15 industrial partners from across Europe; there are over 120 academic researchers involved in NanoCEM work, managing over 60 Ph.D. and Postdoctoral research projects. Industry subscriptions cost between € 30,000 ($ 43,000) and € 60,000 ($ 86,000) per year, based on the company size, and total about € 650,000 ($ 931,000) for NanoCEM each year. Through this organization of research, members of NanoCEM have collective access to a large range of state-of-the-art cemenitious materials equipment, which would otherwise be cost prohibitive. The development concept behind NanoCEM is to have a central platform for academic researchers to share results among themselves and with industry. To this point, there are three main aims of NanoCEM:
- o Research: development of basic research knowledge in the cement and concrete industries
- o Education: prepare the next generation of researchers, especially through placements of university graduates
- o Responsibility: production of solutions that will reduce cost and environmental impact of cement/concrete production

NanoCEM strives to develop a "multidisciplinary integration" of knowledge by focusing on the multifunctionality of research areas; this is a pronounced goal in the consortium governance structure. The NanoCEM Assembly makes decisions on program budgeting for both Core and Partner Projects. The Steering Committee assists by implementing the decisions made, and proposing new budgets and projects. The Industrial Advisory Board identifies research needs and determines the contribution from the industrial partners. The Scientific Committee proposes and prioritizes projects. To date, four core projects are funded solely through NanoCEM. These relate to: (1) hydrate assemblages containing C-S-H;[30] (2) pore structure; (3) organo-aluminate interactions; and 4) reactivity of cemenitious systems. NanoCEM also provides the context for joint projects under the EU's Marie Curie Program. These projects are grouped in four thematic research bundles: (1) deterioration of cement matrices; (2) physical and mechanical verification of performance; (3) new and innovative cement-based materials; and (4) transversal projects.

In order to meet its broader research aims, NanoCEM has put an emphasis on organizing workshops and seminars, sponsoring research in multi-partner projects, and playing a role as a recruitment base for cementitious material researchers. NanoCEM's project success seems to stem from clear roles established for industry and research institutions. It is recognized that industry members provide the funds, while the institutions are responsible for research outcomes. Additionally, through the Marie Currie scheme, PhD students and post-docs are able to obtain experiential positions in industry.

Adoption of projects is carefully considered under the auspices of NanoCEM. There are two consortium meetings a year and during each meeting period stakeholders are separated by industry and academic classifications. In order to be considered, a project area must be proposed by more than one person. NanoCEM management recognizes the importance of diplomacy and democracy in the initial stages of the research process; in order for a project to be fully adopted by NanoCEM, a workshop is held to determine the significance and urgency of the topic under consideration and whether it should be adopted as a NanoCEM-funded and managed project.

There are certain commitments that come with NanoCEM membership. For example, academic research institutions must share some findings from at least one self-financed project in order to compete for NanoCEM funds for other research. The consortium recognizes that isolation in cement/concrete research is one of the greatest challenges in the field. As a meeting place of top researchers in the field, there is encouragement to share information on techniques and research which do not work, since this is what is left out of the literature and is one aspect that contributes to a cumulative waste of research time and funds. At the moment NanoCEM is not accepting new members and is refining its training network. Even though NanoCEM has only been in existence about 5 years, there appears to be a healthy assessment and refinement of its program to ensure success into the future.

### Consortia Comparison
There is merit in comparing VCCTL, SUMMA, ACBM, and NanoCEM, which constitute the leading cement / concrete consortia in the world. In a sense, VCCTL and SUMMA can be paired due to their shared focus on utilizing the ICME approach to develop engineering models of cement/concrete based on sound measurement science. ACBM and NanoCEM are similar to one another due to their shared focus on academic scholarship, fundamental research activities, and the training of students and post-docs.

With respect to the economic benefits of R&D alliances, a NIST Program Office study identifies three distinct benefits to firms collaborating in an R&D alliance: perceptual measures of success (subjective assessment of overall value from industry members), patent measures, and financial measures (revenue or cost savings realized) (Dyer et al., 2006).

---

[30] In cement chemistry notation, C is calcium oxide (CaO), H is water ($H_2O$), and S is silion dioxide ($SiO_2$).

Consortia structure has been studied in detail by Updegrove (1995), from a legal and "social" perspective, to show how they achieve the balance in interests necessary to create outcomes that will fill market needs. He argues that consortia can be powerful forces in standardization, but require as much, if not more, care than SDOs because they are subject to all the influences that impact an SDO as well as their own additional market and stakeholder influences. There are key areas to be considered in the development and upkeep of a consortium, including:.

- Defining the optimal scope of research
- Determining the deliverables that the consortium should create to achieve its mission
- Identifying and segmenting target membership groups whose participation will be needed to not only develop, but enable the adoption, of standards
- Creating attractive member value propositions for each of these target groups to ensure their participation
- Developing an adequate budget and determining appropriate dues structures to support that budget
- Designing a staffing model to serve the needs of the organization
- Adopting key policies to support and add credibility to the organization's technical process

The consortia under consideration are highly differentiated in their scope and goals, as well as the means by which these are set and achieved. Using the NIST and Updegrove criteria as a frame of reference, however, some best practices can be identified.

VCCTL and SUMMA differ in the type of software that they develop, which in turn affects their consortium structure. SUMMA is focused on extending the Stadium software package and has been supported through large-scale contracts with defined start and end dates. There are private and public partners within the SUMMA consortium, but they are all actively bound to the well-defined research goals of the software package. Membership in the VCCTL consortium is heavily based upon obtaining access to the VCCTL software package and its updates; however, members direct other research activities under the auspices of VCCTL. Additionally, the line between VCCTL research and more general HYPERCON research activities becomes somewhat blurred for some projects from time to time. This is especially true of research partnerships that take place between HYPERCON projects and VCCTL members. VCCTL may benefit from a more clear delineation of responsibilities between NIST researchers' and members' research tasks in the formulation of updates to the VCCTL software.

The difference in consortium structure between VCCTL and SUMMA highlights the variation between strategic basic research and more applied research activities. Most obvious is the diminishing willingness of long-time members to invest in VCCTL's strategic basic research activities. It is much easier to define a fixed scope of research in an applied research setting since the basic research platform that is being built upon is already well-defined. Indeed, the head of SUMMA readily expresses his gratitude for VCCTL software results, which has enabled SUMMA's software to progress. Given such a research frame, SUMMA is better poised to determine manageable deliverables and schedules for meeting its research goals. Additionally, consortium models that actively encourage all members to facilitate and add to the actual research, rather than simply direct desired research outcomes, tend to be the most successful at establishing products that are in line with the consortium's specific goals. To date this model is established within SUMMA, NanoCEM, and ACBM much more extensively than within VCCTL.

ACBM and NanoCEM focus on education and formalized academic partnerships more than VCCTL, but their underlying research models are highly focused on policies/structure that supports the organization's technical research processes. The "multidisciplinary integration" noted throughout NanoCEM research modules has connected stakeholders from industry and academia that together have been successful on a number of projects. Additionally, the core component projects which are supplemented by supporting projects facilitates establishment of short term goals and research tasks that can be divided among contributing members. Consortium members obtain an appreciation for other members; this can be achieved, for example, through active research collaborations within the lab or student internships at leading industry companies. Yet, in the NanoCEM model this is reported to be achieved successfully by encouraging academia to fulfill research goals developed by industry members, while highlighting the aspect of direct financing by the industry member. These activities track with the concept of establishing intrinsic membership value, above and beyond strictly financial gains.

During periods of economic downturn, the importance of financing and development of viable budgeting options becomes more important to consortia. A good example of the direct effects of budgeting is evidenced through ACBM's experience over the past twenty years. According to the leadership of ACBM, after the NSF grant funds expired, obtaining full financial support through industry membership has become increasingly challenging. SUMMA and NanoCEM have addressed this issue by highly differentiating membership benefits and fees. This is one approach that VCCTL may want to consider in the future. It tends to dually alleviate the issue of companies not joining because they do not want to pay full membership fees while helping to provide a formalized structure of research responsibilities for each member of the consortium.

When membership is spread among various stakeholder types in the cement/concrete industry, there is likely to be a healthy breadth and scope of research projects/goals suggested by consortium members. Recently the type of stakeholders within VCCTL has become highly concentrated around the chemical admixture industry. Given the experience of other consortia, this may help or hurt VCCTL. It could help VCCTL develop more clarity in its vision, yet it could hurt by unduly limiting the scope and approach that VCCTL is taking to its ICME activities. In either instance, VCCTL management is working to obtain more diverse membership, especially with regards to the DOT community. An educational version of VCCTL scheduled to start development in FY10, known as eVCCTL, will help attract academic stakeholders.

VCCTL leadership has developed some partnerships with the other consortia (through the larger HYPERCON research program) in order to benefit from the research strengths, experiences, and audiences of these other consortia. Most notable is the ACBM/NIST Computer Modeling Workshop, discussed previously in this report. NIST has also partnered with SUMMA members through other consortia, such as the Natural Sciences and Engineering Research Council of Canada (NSERC), Industrial Research Chair on Optimum Maintenance and Durable Repair of Concrete Infrastructure and the DOE-NRC Cement Barrier Partnership.

VCCTL is focused upon using ICME as a strategic basic research methodology in a manner that is not replicated by the other consortia considered in this report. Thus, it should be noted that the models that have worked for other consortia may not be directly applicable to VCCTL. Leaders of these other consortia acknowledge the value of VCCTL activities to their own consortia and the broader cement/concrete industry.

# 6.    Summary

The development of HYPERCON between FY01 and FY09 was tracked using the conceptual framework of *grounded theory*. Grounded theory offers a generalized way to view the evolution of a research program. The use of grounded theory, which encourages continual re-working of linkages between project components and causal relationships, provides a good context for the establishment of the qualitative effects of HYPERCON research. The strategic basic measurement science from HYPERCON and the program history of its VCCTL consortium are both at a stage of development that permits qualitative impact assessment in this manner.

The fact that from FY01 to FY09 HYPERCON's strategic basic research has been at an early stage of grounded theory — identifying  core theoretical concepts, gathering data, and identifying tentative linkages—complicates its assessment and precludes exclusive focus on quantitative indicators of economic impact. There is no viable quantitative metric that can alone track how the results of HYPERCON research that reaches industry, government, and academia is currently being used. Each individual stakeholder (e.g. company, individual) within a stakeholder category applies HYPERCON strategic basic research to their own applied research or production/use activities in a different manner. Thus, the Economics of HYPERCON project interpreted emerging data on quantitative indicators through the lens of a qualitative assessment. Due to the differentiation and scope of the concrete industry, and the often isolated pockets of stakeholders HYPERCON serves, it is most effective to focus on impacts on a single actor/product basis.  This report took that approach, through the use of surveys and case studies. Common themes that emerged from this multi-pronged approach are highlighted in this section.

Due to the OMB regulations on survey distribution, surveys were sent to fewer than 10 people in each stakeholder group.   While these sample sizes are too small to allow for development of econometric models  and decisive quantitative summary statistics, they are adequate for the qualitative approach taken and in some cases (small stakeholder groups) even quite representative. Care was taken in development of survey questions in order to garner the most representative and valuable feedback. The identification of participants was done through a careful process in order to obtain information from those most familiar with the theory and applications being enabled by  HYPERCON research.

This Economics of HYPERCON study was designed to provide BFRL managers with a tool to review HYPERCON program activities and impacts to date. While the program is evaluated retrospectively, some findings are particularly relevant in the context of future research directions.  Section 6.1 summarizes retrospective impacts of HYPERCON, while section 6.2 looks ahead to HYPERCON's future.


6.1 HYPERCON Retrospective Economic Impacts

As grounded theory suggests, HYPERCON technical focus areas have evolved over the period from FY01 through FY09, building upon research discoveries from period to period and adapting to changing industry needs.  While fundamental measurement science issues remain in each of the five HYPERCON technical areas, HYPERCON's involvement in these areas since FY01 has generally evolved from providing the underlying measurement science toward addressing focused measurement issues enabling technical problem-solving and technology transfer. The cement/concrete industry is widespread and consists of many stakeholder groups. Their interests vary greatly, as does market share. These factors pose challenges to both HYPERCON research managers—tasked with *meeting the most important concrete performance predictions needs of U.S. industry in the most cost-effective manner*—and any economic impact assessment of the research program.. Yet, the process adopted does allow for a deep understanding of the value of HYPERCON research to relevant stakeholder groups.

To the extent possible, this study included FY09 HYPERCON activities to account for major changes in program goals between FY08 and FY09. Yet, during the survey process it was clear that stakeholders had not recognized tangible outcomes from this change; as should be expected, it will take years to develop and deliver new outcomes. Notably, in FY09 HYPERCON was tied much more strongly and explicitly to national documents outlining the need for the kind of research proposed. The research approach was directly identified as ICME for the first time. The Materials Characterization project was revamped to put the major emphasis on fly ash research. Modeling cement paste rheology with fly ash was assigned to the Rheology project. Additionally, long-range milestones were added to the Hydration Modeling project for fly ash and slag modeling. The interest of academic stakeholders in HYPERCON activities was evaluated through exploration of both formal and informal collaborations and knowledge sharing. Interest appears to be quite strong and growing, with informal academic collaborations being an important venue for knowledge exchange. The ACBM/NIST Computer Modeling Workshop, which attracts a variety of stakeholders ranging from the academic to the industrial communities, has in recent years attracted significant interest among those specifically focused on computer modeling techniques, often times with no direct relationship to cement/concrete research. While this is a sign of HYPERCON's leading role in the ICME community at large, survey results indicate that the workshop is struggling to meet the growing, multi-faceted expectations of its participants. The Electronic Monograph, an inclusive record of HYPERCON research findings, is a relevant tool that serves many stakeholder groups' interests. The Monograph has proven to be one of the most effective (of many) channels for HYPERCON's impact within the academic realm. The HYPERCON Program's strong overall H-Factor from FY01 to FY09 points to the program's value and scholarly respect. Its publication base is large and growing steadily, with a clear upward trend in the number of attributed citations. One academic stakeholder expressed the sentiments of many when responding:

> *The NIST concrete group has put concrete at equal footing with other engineered materials, namely metals and ceramics. NIST thus largely contributes to reinventing concrete science and engineering as an academic discipline in the U.S. and worldwide.*

In the standards arena, there was an increasing trend in SRM sales from FY01 to FY09, particularly to foreign customers. HYPERCON's success in voluntary consensus standards development is more difficult to track, in part because of the very nature of standards development as a team exercise. Individuals and entities for the most part cannot claim "ownership" of a standard. While all of HYPERCON's clear contributions and direct impacts on standards from FY01 to FY09 are not evident, researchers' leadership positions on SDO committees are a positive indicator of future impact. Furthermore, SDO stakeholders generally support HYPERCON research and expect it will inform their future standards development needs.

As expected for a strategic basic research program, as the view shifts from academia to industry, there is a general trend for HYPERCON's retrospective impacts to become more diffuse. Stakeholder groups become more diverse in their levels of technical expertise and computer savvy and consequently, in how they view and plan to use HYPERCON research. Some of those interviewed are generally concerned with keeping up with the competition, while others are striving to position themselves as industry leaders. This diversity creates a challenge to HYPERCON to create a single mechanism for transfer of its research to industry and also affects the level of willingness for some companies to share their own strategic basic research activities with HYPERCON researchers. This is a common issue faced by any strategic basic research program and it is expected that as HYPERCON's grounded theory consolidates in the future and yields more tangible results, a more formalized solution to this problem may emerge.

The VCCTL is currently faced with an economy-driven decline in consortium membership, together with a consolidation towards exclusive membership among a lone stakeholder group (chemical admixtures). While consortium membership has fluctuated, past members generally found value in their participation. Past members support the idea of VCCTL research, but are finding it difficult to justify membership fees on a long-term basis. Among current and past consortium members and non-members alike, there appears to be interest in the VCCTL software, yet it does not align well with actual software use. Rather than a limitation in the VCCTL software itself (members report substantial improvement in the software interface), this likely highlights the naissance of the ICME approach, for which VCCTL is a leader. While many see a "significant potential application" for VCCTL in their future work, there is also a growing impatience for more tangible results.

The case studies of relevant cement/concrete consortia highlighted some areas in which VCCTL is a leader as well as some comparative disadvantages in research structure. Particularly, the differences in consortia structure highlight the difference between strategic basic research and more applied research activities. The largest and most diverse cement/concrete consortium, NanoCEM, which is also the newest of the four reviewed, seems to benefit highly from a strict structure of research responsibilities and roles put in place by consortium management. Also, it is clear that the most widespread and quickest results occur when there is a wide pool of stakeholder groups represented by the consortium membership. In all consortia it is evident that members expect tangible results in return for financial contributions. This is much more difficult to achieve for a strategic basic research program, like VCCTL, than for consortia conducting applied research.

Across all stakeholders there is consistently a high respect for HYPERCON researchers and its program leader. Though, especially in industry, there is some frustration over the speed at which findings are produced for HYPERCON research projects. This is a hallmark issue for strategic basic research programs. The usefulness and timing of their research findings are more difficult to anticipate, requiring flexibility in program planning.

HYPERCON is making progress in identifying tentative linkages among its core theoretical concepts. This is clear when questions about HYPERCON technical areas are asked to multiple stakeholder groups interested in a common area. Generally, it was found that while many of those surveyed are not aware of the details of HYPERCON research, those that were aware are confident that further HYPERCON research will be applicable to their work, particularly in those technical areas that are furthest along. ICME-related rheology is the exception, enjoying both good awareness and support among stakeholders. X-ray diffraction was found to be highly relevant to a range of stakeholders, from the beginning to the end of the cement/concrete industry supply chain. HYPERCON's P2P research—like P2P research in most areas—is struggling to define HYPERCON's role in fulfilling industry needs. Stakeholder groups were unanimous in looking forward to using the results of further HYPERCON research in their work. HYPERCON's progress in making linkages is also evident in its success at developing venues for bringing stakeholders together. It is important for stakeholders to have buy-in to HYPERCON's ultimate goal--in grounded theory's terms, conceptually dense theory for cement/concrete performance prediction. At this point, it is a matter of meeting HYPERCON expectations, defined internal to NIST and externally, and delivering the performance prediction models and tools its stakeholders clearly want.

While the nature of current HYPERCON research is still quite diffuse, there is great potential for significant future impact through multiple channels and serving multiple interests. There is no doubt that given the current economic climate there is a challenging road ahead for all cement/concrete strategic basic research. In the case of HYPERCON, success may depend on maintaining a delicate balance between being too ambitious and too weak in the promised outputs; it is key to avoid disenchanting those stakeholders that currently are actively engaged and highly supportive of HYPERCON's efforts. While stakeholder awareness of HYPERCON activities is not strong in some areas, this is not necessarily a place to focus on improvement in the immediate term. A more productive effort may be to develop and execute a vision that consolidates the overall program of research, so that when HYPERCON communicates with the currently uniformed, it can do so in terms of a compelling business case of mutual benefit to HYPERCON and U.S. cement/concrete industry.

## 6.2 HYPERCON: Looking Forward

Given the potential growth and importance of the industry in the coming years, NIST is poised to play an increasingly important role in cement and concrete research. The combination of sustained high asphalt prices, challenging U.S. economic conditions, and the need to improve and expand highway infrastructure creates favorable conditions for continued increases in concrete highway paving. Additionally, over the past decade advances have been made in the use of waste materials such as coal fly ash and blast furnace slag in cements, of crushed glass products in aggregates, and of recycled concrete. This is a trend that will continue and those surveyed expressed interest in understanding the chemistry and kinetics of concrete containing recyclables, especially through ICME interfaces. The combination of the current economic slowdown and environmental awareness contributes to a growing need to explore the value of waste stream materials to the concrete industry and construction industries as a whole. The use of such waste stream materials requires research, including measurement science and materials characterization, to be used effectively as a substitute for virgin materials in cement and concrete production.

As discussed above, there is a need for HYPERCON to consolidate its research projects into a program with a clear and succinct vision that is readily communicated. The grounded theory approach taken here shows that HYPERCON has proven its leadership in strategic basic research for cement/concrete. Wider dissemination of this expertise may in turn attract a wider user base to the VCCTL, both as potential members and collaborators. The VCCTL membership structure and governance processes may benefit from a review of the methods employed by other cement/concrete consortia.

HYPERCON research is applied throughout academia and industry in diverse and highly meaningful ways, which at this point can not be accurately tracked solely through monetary terms. Once the strategic basic research produced through HYPERCON permeates the industry and cost savings are apparent on an aggregate level, a review based on quantitative economic metrics should be possible. Then, the approach taken in this study could be enhanced with an econometric model based on two periods of data on HYPERCON quantitative and qualitative metrics to compare and use in the modeling.

This is the first economic impact study of a NIST program using grounded theory. Some lessons were learned that will help refine future studies, not only of HYPERCON, but of other strategic basic research programs in their naissance.

In order to have robust statistics to create an econometric model of impact, more survey participants than in this study are needed in each stakeholder group. When larger sample sizes are not possible, though, surveys can still be a valuable tool to inform impact studies if properly designed and administered. In this study, it was critical to work with HYPERCON technical experts to methodically and carefully design survey questions and to determine candidates for survey participation. It was also very useful to ask common questions to multiple stakeholder groups; similar responses from different perspectives serve to strengthen the qualitative assessment.

The case studies were useful as a means of determining the position that the VCCTL takes in the cement/concrete research consortia community. Ideally, interviews would take place of companies in the industry which currently do or have in the past held membership in more than one of the consortia under consideration. This type of interview process was initially planned for this study; however, few companies hold multiple consortia memberships and those which do were reluctant to provide direct comparison feedback. Additionally, determining the true operating budgets of the other consortia was not possible due to a lack of willingness by leadership to provide the information. Yet, in general the detailed descriptions of other cement/concrete consortia proved useful in assessing alternative consortia governance styles.

This study has used grounded theory to assess HYPERCON performance from FY01 through FY09. Assessment of qualitative and quantitative indicators demonstrates that HYPERCON plays an important and potentially growing role in cement/concrete strategic basic research.

# References

*ASCE State of the US Infrastructure* (2005) Available at: http://www.asce.org/.

*Assessment of Department of Defense Basic Research: Appendix D: Definitions of Basic, Applied, and Fundamental Research.* National Research Council Publication. (2005)

Bateman, I.J., Jones, A.P, Jude, S. and Day, B.H. (2006) Reducing gains/loss asymmetry: A virtual reality choice experiment (VRCE) valuing land use change. CSERGE Working Paper EDM 06-16. Norwich, UK: University of East Anglia.

Concrete Durability: A Multibillion-Dollar Opportunity - National Material Advisory Board, National Research Council (1987) NMAB-437.

*Concrete Infocus.* (Winter 2008) National Ready Mixed Concrete Association.

Corbin, J. and A. Strauss. (2008) *Basics of Qualitative Research.* Los Angeles, CA: Sage Publications.

Dohmen, T., Falk, A., Huffman, D., Sunde, U., Schupp, J., and Wagner, G.G. (2005) Individual Risk Attitudes: New Evidence from a Large, Representative, Experimentally-validated Survey. IZA Discussion Paper 1730. Bonn, Germany: Institute for the Study of Labor.

Dyer, J. et al. "Determinants of Success in R&D Alliances." (August 2006) NISTIR 7323: http://www.atp.nist.gov/eao/ir-7323/ir-7323.pdf.

Eidt, C. and R. Cohen. (January–February 1997) "Reinventing Industrial Basic Research," Research Technology Management: 29–36.

Eldridge, L.P. (February 1999) "How Price indexes affect BLS productivity measures." Monthly Labor Review: 35-46.

Finkelstein, L. *Theory and philosophy of measurement.* In: *Handbook of Measurement Science* (1982) Vol. 1, John Wiley, Chichester, pp. 1–30.

Gallaher, M. et al. "Cost Analysis of Inadequate Interoperability in the U.S. Capital Facilities Industry;" (August 2004) NIST GCR 04-867.

Glaser, B.G. and Strauss, A.L. (1967) *The discovery of grounded theory; strategies for qualitative research.* Chicago, IL: Aldine Pub. Co.

Greene, W. (2008) *Econometric Analysis.* New Jersey: Pearson, Prentice Hall.

*High-Performance Construction Materials and Systems: An Essential Program for America and Its Infrastructure,* (1993) ASCE Tech. Rpt. 93-5011.

Hirsch, J. E. (2005). "An index to quantify an individual's scientific research output". *PNAS* 102 (46): 16569–16572.

Holton, *W.C. Power Surge: Renewed Interest in Nuclear Energy*. (November 2005) Environmental Health Perspectives. Vol. 113: Number 11.

*Integrated Computational Materials Engineering: A Transformational Discipline for Improved Competitiveness and National Security* (National Research Council, National Academies Press, in press, 2008)

Martin, B. and P. Tang. (June 2007) "The benefits from publicly funded research." SPRU Electronic Working Paper Series Issue *Paper Number 161*: http://www.sussex.ac.uk/spru/documents/sewp161.pdf.

Michell, J. (1986), Measurement Scales and Statistics: A Clash of Paradigms, *Psychological Bulletin,* 100:3, pp. 398-407.

National Science Board Statement, *In Support of Basic Research*, NSB-93-127 (14 May 1993) http://www.nsf.gov/nsb/documents/1993/nsb93127/nsb93127.htm

*National Science Board Commission on the Future of the National A Foundation for the 21st Century: A Progressive Framework for the National Science Foundation.* (November 20, 1992) Washington, D.C. http://www.nsf.gov/pubs/stis1992/nsb92196/nsb92196.txt

National Science Board. *A Companion to Science and Engineering Indicators 2008.* (2008) Available at: http://www.nsf.gov/statistics/nsb0803/start.htm.

National Research Council of the National Academies. *Integrated Computational Materials Engineering: A Transformational Discipline for Improved Competitiveness and National Security.* (2008) National Academies Press, Washington, D.C.

Pavitt, K. *National Policies for Technical Change. What are the increasing returns to economic research?* (November 1996) Proc. Natl. Acad. Sci. Vol. 93, p 12693-12700. Available at: http://www.pnas.org/cgi/reprint/93/23/12693

Perlman, A. PCA Perspectives: Paving the Way for Economic Recovery. (2009) Available at: http://www.cement.org/Paving%20The%20Way%20For%20Economic%20Recovery%20LR%205.4.pdf.

Ruegg, R. and I. Feller. *A Toolkit for Evaluating Public R&D Investment: Models, Methods, and Findings from ATP's First Decade.* (2003) NIST GRC 03-857, prepared for Economic Assessment Office.

*Roadmap 2030: The U.S. Concrete Industry Technology Roadmap* (v1.0), (2002) ACI Strategic Development Council.

Stevens, S.S. "On the Theory of Scales and Measurement," (June 1946) *Science*, Vol. 103, no. 2684.

Stokes, D. *Pasteur's Quandrant: Basic Science and Technological Innovation.* (1997) Washington, D.C.: Brookings Institution Press.

Sullivan, E. "Update: Paving, the New Realities." (July 2009), prepared for the Portland Cement Association (PCA). Available at: http://www.cement.org/asphaltreport-July%202009.pdf.

Tassey, G. *R&D Policy Models and Data Needs,* NIST Program Office; APPAM 1999 Research Conference. (Nov 4, 1999)

*Technology for America's Economic Growth A New Direction to Build Economic Strength.* Washington, DC, (1993) White House Office of the Press Secretary. http://www.itsdocs.fhwa.dot.gov/JPODOCS/BRIEFING/7423.pdf

*The Concrete Pavement Roadmap: Long-Term Plan for Concrete Pavement Research and Technology*, FHWA. (2006) Available at: http://www.fhwa.dot.gov/pavement/pccp/pubs/05047/.

*Unlocking Our Future: Toward a New National Science Policy.* (September 24, 1998) A report to Congress by the House Committee on Science.

Updegrove, A. (1995) "Setting Standards on Firm Foundations: Consortia Structures and Consensus Building." *Geo Info Systems*: Vol. 5, No. 5.

USGS Cement Statistics. (Dec. 2, 2007) Available at: http://minerals.usgs.gov/ds/2005/140/cement.pdf.

Yin, R. K. (1984) *Case Study Research, Design and Methods.* Sage Publications, Newbury Park.

www.ingramcontent.com/pod-product-compliance
Lightning Source LLC
Chambersburg PA
CBHW080305180526
45167CB00006B/2681